My Book

This book belongs to

Name:_____

Copy right © 2019 MATH-KNOTS LLC

All rights reserved, no part of this publication may be reproduced, stored in any system or transmitted in any form, or by any means, electronic, mechanical, photocopying, recording, or otherwise without the written permission of MATH-KNOTS LLC.

Cover Design by :
Gowri Vemuri

First Edition :
April, 2020

Second Edition :

January, 2021

Author :
Gowri Vemuri

Editor :
Ritvik Pothapragada

Questions: mathknots.help@gmail.com

Dedication

This book is dedicated to:
My Mom, who is my best critic, guide and supporter.
To what I am today, and what I am going to become tomorrow,
is all because of your blessings, unconditional affection and support.

This book is dedicated to the
strongest women of my life,
my dearest mom
and
to all those moms in this universe.

G.V.

INDEX

Notes	9 - 20
Addition	21 - 37
Subtraction	38 - 52
Mixed : Addition and Subtraction	53 - 61
Multiplication	62 - 76
Division	77 - 86
Rounding the whole numbers	87 - 96
Write each numeral in words	97 - 109
Write the numerals	110 - 130
Place values	131 - 145
Round the whole numbers	146 - 158
Answer Keys	159 - 194

WHOLE NUMBER

Notes

Whole Numbers Notes

The word standard means regular. Numbers in standard form are whole numbers or natural numbers.

Example : The number "nine hundred seventy two" in standard form is 972.

Real number system

Irrational numbers
$\sqrt{2}, \sqrt{3}, \sqrt{5}, ...$

Rational numbers

Integers

Whole numbers

Natural numbers
1, 2, ...

0, 1, 2, 3...

..., -3, -2, -1, 0, 1, 2, 3, ...

$-\infty ..... -2, \frac{-3}{2}, -1, \frac{-1}{2}, \frac{-1}{3}, 0, \frac{1}{3}, \frac{1}{2}, 1, \frac{3}{2}, 2 ... \infty$

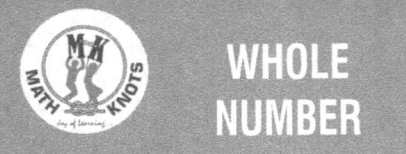

WHOLE NUMBER

Notes

Number spellings :

1 : one	21 : twenty - one	1,000 : one thousand
2 : two	22 : twenty - two	2,000 : two thousand
3 : three	23 : twenty - three	100,000 : one hundred thousand
4 : four	24 : twenty - four	1,000,000 : one million
5 : five	25 : twenty - five	10,000,000 : ten million
6 : six	26 : twenty - six	100,000,000 : one hundred million
7 : seven	27 : twenty - seven	2,000 : two thousand
8 : eight	28 : twenty - eight	32,000 : thirty - two thousand
9 : nine	29 : twenty - nine	200,000 : two hundred thousand
10 : ten	30 : thirty	2,000,000 : two million
11 : eleven	40 : forty	20,000,000 : twenty million
12 : twelve	50 : fifty	200,000,000 : two hundred million
13 : thirteen	60 : sixty	
14 : fourteen	70 : seventy	
15 : fifteen	80 : eighty	
16 : sixteen	90 : ninety	
17 : seventeen	100 : one hundred	
18 : eighteen		
19 : nineteen		
20 : twenty		

"ty" rule : When we are connecting two numbers and the first ends in "ty", we use a hyphen between the words.

Place Value of digits and Expanded Form:

7,286,510

A whole number consists of digits, each with a certain place value. Whole numbers do not include fractions or decimals.

The number 784 has 7 in the hundreds place, 8 in the tens place, and 4 in the ones place.

Each digit in the number 784 has a value. The value of 7 is 700, the value of 8 is 80, and the value of 4 is 4.

WHOLE NUMBER

Notes

Rounding whole numbers :

The purpose of rounding is to provide estimated values (Approximate values).
When rounding, we drop a few numbers and get an approximated value.
Rounding involves losing some accuracy in the data.

Example :   If 27,594 people attend a broad way show, the news may report that there were approximately 30,000 people at the game. 30,000 is 27,594 rounded to "the nearest ten thousands place."

Steps to round :

1. Identify the digit we want to round the value to.
2. Look at the digit to the right of the place value we want to round to.
3. If the digit to the right is 5 or over, add 1 to the digit we want to round to.
4. If the digit to the right is 4 or under, the digit remains the same.
5. Remember : The places to the right of the hundreds place are filled with zeros.

In other words:

> **FIVE AND ABOVE, GIVE IT A SHOVE.**
> **FOUR AND BELOW, LEAVE IT ALONE.**

Commutative Property of Addition:

Addition is commutative: The order of addition does not alter the value.

Examples:

9 + 10 = 10 + 9
33 + 44 = 44 + 33
27,500 + 99,800 = 99,800 + 27,500

Associative Property of Addition:

Addition is associative: The order of addition does not matter when the numbers are grouped in different ways. The result of the addition remains the same.
In other words: "The order in which we add a group of numbers does not alter its value.

Example:
5 + (11 + 16) = (5 + 11) + 16

WHOLE NUMBER

Notes

Properties of Addition:

Identity Property of Addition:
Adding 0 to any number does not alter its value. Hence we call 0 as additive identity.

Example :

75 + 0 = 75

Adding Large Numbers

To add large numbers, it is important to line up the digits according to place values.
DON'T FORGET TO ADD THE NUMBER YOU CARRIED OVER!

Example :
```
   5,105
 +3,978
```

Adding Three Numbers and more :

To add three large numbers, we must always line up numbers according to their place value. Use commas to make the numbers easier to read.
DON'T FORGET TO ADD THE NUMBER YOU CARRIED OVER!

Example : 22,322 + 10,101 + 392,525

Subtracting numbers :

The rules for subtracting numbers up to millions are similar to adding 3 digit numbers,
Align the numbers according to the place values of the numbers.

Example :   The millions, hundred thousands, ten thousands, and thousands columns are used.

The key to subtraction is borrowing. If we can not subtract two digits, then we must borrow 1 from the next digit to the right making it equal to 10, add 10 to the digit value from where you are borrowing.

Multiplication :

Multiplication is a "repeated addition".

Properties of multiplication:

Commutative is commutative :
Multiplication is commutative: The order of multiplication does not alter the value.

Examples :

100 X 7 = 7 X 100
25 X 20 = 20 X 25

Associative property :

Multiplication is associative: The order of Multiplication does not matter when the numbers are grouped in different ways. The result of the Multiplication remains the same.
In other words: "The order in which we multiply a group of numbers does not alter its value.

Example :

5 X (7 X 11) = (5 X 7) X 11

The multiplicative identity :

Multiplying 1 with any number does not alter the value of the number. Hence we call 1 as multiplicative identity.

Examples :

37 X 1 = 37

48,932 X 1 = 48,932

Distributive property:

The distributive property of multiplication : multiplication may be distributive over addition or subtraction.

Examples :

21 X (15 + 7) = (21 X 15) + (21 X 7)

30 X (20 - 9) = (30 X 20) - (30 X 9)

The zero property of multiplication:

Multiplying any number by 0 equals the product value to 0.

Examples :

27 X 0 = 0
0 X 93,905 = 0

Multiplying a 2 - digit number by a 1 - digit number:

Example :

$$\begin{array}{r} 3\,9 \\ \times \quad 4 \end{array}$$

1. First multiply 4 by the ones place of the top number.
2. Write the digit in the ones place of that product in the ones place of the answer.
3. Carry over the digit in the tens place above the 3 (which is in the tens place of the top number).
4. Multiply the 4 by the digit in the tens place of the top number and then add the number that you carried to this product.
5. Write the number in front of the ones place in the answer.

Multiplying a 3 - digit number by a 1 - digit number:

Example :

$$\begin{array}{r} 1\,2\,7 \\ \times \quad 8 \end{array}$$

1. First multiply 8 by the ones place of the top number.
2. Write the digit in the ones place of that product in the ones place of the answer.
3. Carry over the digit in the tens place of the product above.
4. Multiply 8 by the digit in the tens place of the top number and then add the number that you carried to this product.
5. Write the digit in the ones place of the product in the tens place of the answer.
6. Carry over the digit in the tens place of the product above the 1 (which is in the hundreds place of the top number).
7. Multiply the 8 by the digit in the hundreds place of the top number and then add the number that you carried that you carried to this product.
8. Write the number in front of the tens place in the answer.

"When setting up multiplication problems, always write the larger number on the top"

WHOLE NUMBER

Notes

Multiplying a 2 - digit number by a 2 - digit number :

Steps for multiplying two 2 digit number :

$$\begin{array}{r} 17 \\ \times\ 14 \end{array}$$

1. Multiply the top number by the one digits in the bottom number.

2. Place a zero under the ones digit of the product (we are now multiplying by the tens place, so we must have a placeholder (X) for the ones place. Multiply the top number by the tens digit of the bottom number.

3. Add the two products together.

Multiplying a 3 - digit number by a 2 - digit number:

Example :

$$\begin{array}{r} 325 \\ \times\ \ \ \ 15 \end{array}$$

1. Multiply the top number by the ones digits in the bottom number. Remember to "carry over" digits in the "tens place" of every product

2. Place a zero under the ones digit of the product (we are now multiplying by the tens place, so we must have a place holder (X) for the ones place). Multiply the top number by the tens digit of the bottom number.

3. Add the two products together.

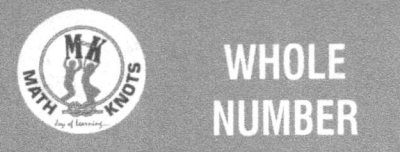

WHOLE NUMBER

Multiplying a 3 - digit number by a 3 - digit number:

Example :

$$\begin{array}{r} 234 \\ \times\ 125 \end{array}$$

1. Multiply the top number by the ones digits in the bottom number.
2. Place a zero under the ones digit of the product (we are now multiplying by the tens place, so we must have a place holder). Multiply the top number by the tens digit of the bottom number.
3. Place a value X under the ones digit and the tens digit of the product (we are now multiplying by the hundreds place). Multiply the top number by the hundreds digit of the bottom number.
4. Add the three products together.

Division with single digit factors :

Division is the inverse operation to multiplication. It is also known as repeated subtraction.

If 42 ÷ 6 = 7, then 7 X 6 = 42 and 6 X 7 = 42.

Division : "How many groups of divisor fits into dividend ?"

24 ÷ 6 means, how many groups of 6 fit into 24 ? OR what number times 6 is 24 ?

Parts of a division :

Dividend : the number being divided
Divisor : the number that divides

Quotient : The answer to a division problem. The number of times the divisor goes into the dividend; the answer to a division problem.

$$81 \div 3 = 27$$
Quotient   Divisor   Dividend

Division with single digit factors :

There are some quick ways to determine if a given number is divisible by another number.

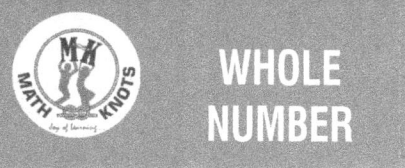

WHOLE NUMBER

Divisibility rules :

Divisible by 2 : If the number is even, then it is divisible by 2. [Ends in 0 , 2 , 4 , 6 , or 8]

Divisible by 3 : If the sum of the number's digits is divisible by 3, then it is divisible by 3.
Example : 369   3 + 6 + 9 = 18 ; 18 is divisible by 3

Divisible by 4 : If the last two digits of the number are divisible by 4.
Example :   412,424 is divisible by 4. Since 24 is divisible by 4.

Divisible by 5 : If the number's last digit is a 0 or 5, then it is divisible by 5.

Divisible by 6 : If the number is divisible by 2 and 3, then it is divisible by 6.

Divisible by 9 : If the sum of the number's digits is divisible by 9, then it is divisible by 9.
              (Similar to the divisible by 3 rule)

Divisible by 10 : If the number ends in 0, then it is divisible by 10.

Long Division :

Long division is a way of dividing numbers that are too large for us to do quickly.
A key thing to remember with long division is to always line up your digits properly.

Example :

Right   **Wrong**

3 goes into 12 four times. Always line up the quotient with the corresponding place value

$$3\overline{)12} = 4$$

3 does not divide into 2 four times. When it is written like this, the 4 is lined up with the 2, not 1; therefore it is communicating the wrong thing.

$$3\overline{)12} = 4$$

WHOLE NUMBER

There are 4 steps to long division :

| #1 Divide |
| #2 Multiply |
| #3 Subtract |
| #4 Check |
| #5 Bring down |

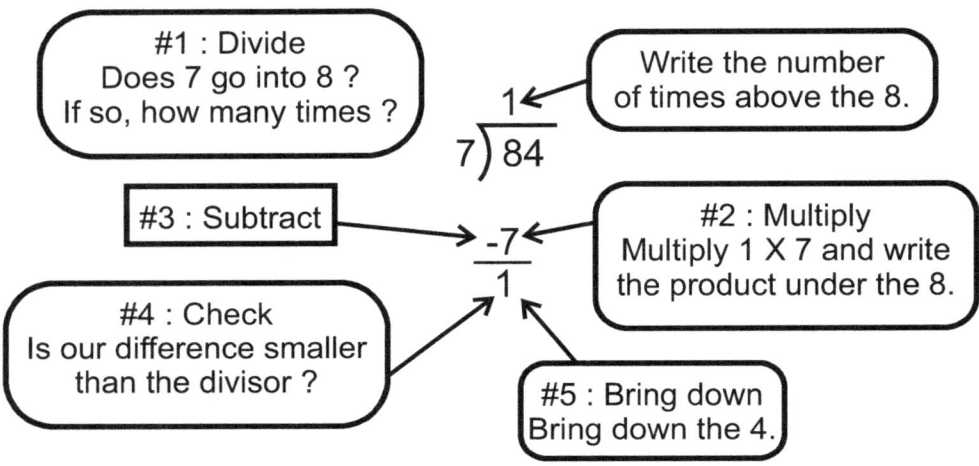

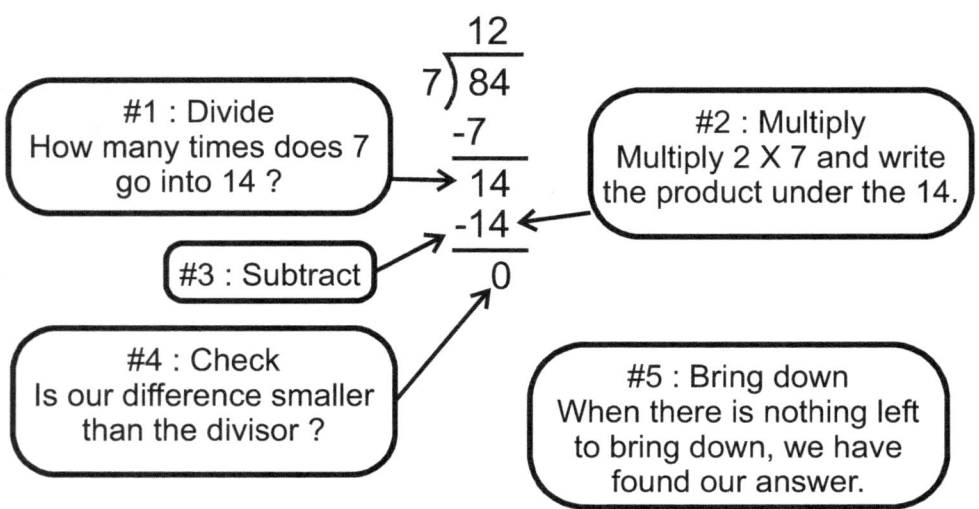

WHOLE NUMBER

Long Division : 1 Digit divisor , 3 digit dividend

For three - digit dividends, we follow the same method: Divide, multiply, subtract, check, bring down.

NOTE : Since 5 doesn't go into 3, so an x should be written as a place holder. Now expand the digits to 37.

How many 5s fit into 37 ?

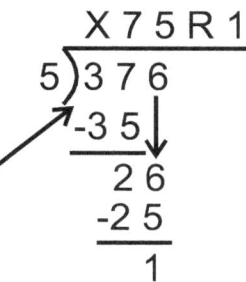

NOTE : After subtracting, bring down, the next number to divide 5 by 26. How many 5's divide into 26 ?

When we multiply and subtract again, we get a remainder of 1.

Dividing larger number by 2 - digit numbers

For three - digit dividends, we follow the same method: Divide, multiply, subtract, check, bring down.

We always start with the highest place value when dividing.

Remember the 4 steps
1. Divide
2. Multiply
3. Subtract
4. Check
5. Bring Down

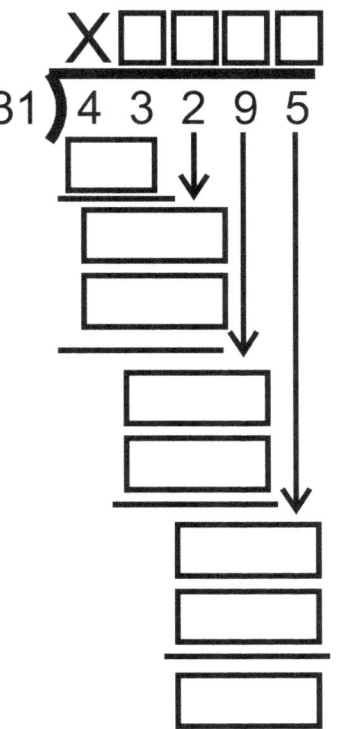

31 does not fit into 4, it does fits into 43. Start writing the quotient above 3.

Whole Numbers

Find the sum of the numbers given below

(1) 208 + 446

(2) 201 + 223

(3) 212 + 485

(4) 22 + 297

(5) 369 + 111

(6) 118 + 62

(7) 172 + 364

(8) 496 + 165

(9) 367 + 132

(10) 103 + 333

WHOLE NUMBERS

Basic Math

Find the sum of the numbers given below

(11) 156 + 170

(12) 272 + 12

(13) 19 + 195

(14) 365 + 72

(15) 314 + 177

(16) 190 + 433

(17) 403 + 383

(18) 330 + 64

(19) 167 + 215

(20) 236 + 271

WHOLE NUMBERS

Basic Math

Find the sum of the numbers given below

(21) 185 + 102

(22) 471 + 212

(23) 111 + 258

(24) 173 + 467

(25) 433 + 372

(26) 433 + 99

(27) 1 + 296

(28) 158 + 348

(29) 361 + 179

(30) 120 + 115

WHOLE NUMBERS

Basic Math

Find the sum of the numbers given below

(31) 110 + 342

(32) 324 + 6

(33) 155 + 160

(34) 173 + 73

(35) 354 + 233

(36) 169 + 493

(37) 395 + 490

(38) 174 + 432

(39) 294 + 63

(40) 42 + 366

WHOLE NUMBERS

Basic Math

Find the sum of the numbers given below

(41) 425 + 221

(42) 429 + 218

(43) 279 + 118

(44) 282 + 10

(45) 473 + 452

(46) 99 + 458

(47) 55 + 396

(48) 196 + 105

(49) 276 + 205

(50) 478 + 253

WHOLE NUMBERS

Basic Math

Find the sum of the numbers given below

(51) 360 + 913 + 17

(52) 169 + 970 + 571

(53) 141 + 613 + 847

(54) 118 + 900 + 385

(55) 197 + 584 + 789

(56) 124 + 889 + 981

(57) 779 + 344 + 718

(58) 143 + 924 + 51

(59) 88 + 867 + 456

(60) 751 + 807 + 606

WHOLE NUMBERS

Basic Math

Find the sum of the numbers given below

(61) 968 + 459 + 496

(62) 956 + 629 + 650

(63) 58 + 646 + 959

(64) 736 + 149 + 213

(65) 582 + 449 + 273

(66) 329 + 952 + 482

(67) 700 + 251 + 881

(68) 906 + 61 + 586

(69) 689 + 809 + 16

(70) 443 + 44 + 499

WHOLE NUMBERS

Find the sum of the numbers given below

(71) 88 + 885 + 306

(72) 542 + 936 + 252

(73) 638 + 445 + 254

(74) 419 + 236 + 205

(75) 797 + 215 + 629

(76) 777 + 431 + 843

(77) 521 + 660 + 526

(78) 301 + 125 + 200

(79) 605 + 319 + 238

(80) 762 + 975 + 532

WHOLE NUMBERS

Basic Math

Find the sum of the numbers given below

(81) 368 + 180 + 977

(82) 926 + 762 + 975

(83) 152 + 378 + 1

(84) 971 + 720 + 811

(85) 928 + 861 + 998

(86) 473 + 483 + 484

(87) 186 + 356 + 977

(88) 690 + 353 + 383

(89) 725 + 769 + 643

(90) 232 + 326 + 257

WHOLE NUMBERS

Basic Math

Find the sum of the numbers given below

(91) 204 + 742 + 478

(92) 518 + 537 + 217

(93) 780 + 3 + 921

(94) 833 + 191 + 459

(95) 878 + 207 + 808

(96) 275 + 894 + 143

(97) 65 + 308 + 513

(98) 704 + 545 + 445

(99) 569 + 387 + 291

(100) 189 + 26 + 654

WHOLE NUMBERS

Basic Math

Find the sum of the numbers given below

(101) 941 + 265 + 469 + 168

(102) 523 + 542 + 249 + 558

(103) 880 + 687 + 848 + 725

(104) 621 + 619 + 496 + 492

(105) 774 + 287 + 705 + 677

(106) 132 + 315 + 499 + 518

(107) 416 + 689 + 5 + 905

(108) 31 + 975 + 558 + 154

(109) 64 + 700 + 451 + 155

(110) 921 + 99 + 759 + 259

WHOLE NUMBERS

Basic Math

Find the sum of the numbers given below

(111) 859 + 881 + 183 + 103

(112) 284 + 132 + 266 + 252

(113) 57 + 386 + 767 + 760

(114) 940 + 302 + 566 + 949

(115) 760 + 493 + 703 + 815

(116) 781 + 284 + 574 + 362

(117) 913 + 181 + 590 + 915

(118) 338 + 919 + 717 + 670

(119) 883 + 599 + 335 + 311

(120) 76 + 845 + 890 + 648

WHOLE NUMBERS

Find the sum of the numbers given below

(121) 912 + 256 + 937 + 946

(122) 406 + 212 + 918 + 354

(123) 134 + 836 + 651 + 962

(124) 24 + 633 + 470 + 698

(125) 521 + 576 + 320 + 702

(126) 688 + 411 + 277 + 670

(127) 870 + 165 + 888 + 992

(128) 272 + 332 + 35 + 76

(129) 870 + 918 + 824 + 573

(130) 326 + 244 + 9 + 950

WHOLE NUMBERS

Basic Math

Find the sum of the numbers given below

(131) 682 + 945 + 772 + 814

(132) 866 + 770 + 806 + 815

(133) 856 + 31 + 630 + 599

(134) 541 + 57 + 165 + 708

(135) 841 + 103 + 102 + 21

(136) 521 + 518 + 879 + 825

(137) 201 + 583 + 100 + 109

(138) 300 + 889 + 504 + 132

(139) 624 + 543 + 835 + 113

(140) 452 + 372 + 753 + 925

WHOLE NUMBERS

Basic Math

Find the sum of the numbers given below

(141) 438 + 196 + 618 + 425

(142) 101 + 741 + 229 + 48

(143) 998 + 859 + 603 + 94

(144) 911 + 580 + 841 + 798

(145) 321 + 85 + 626 + 466

(146) 340 + 946 + 212 + 174

(147) 827 + 301 + 265 + 110

(148) 542 + 382 + 833 + 361

(149) 992 + 642 + 162 + 840

(150) 817 + 270 + 423 + 799

WHOLE NUMBERS

Find the difference of the numbers given below.

(151) 59 − 0

(152) 73 − 36

(153) 73 − 32

(154) 60 − 23

(155) 32 − 22

(156) 72 − 23

(157) 33 − 3

(158) 67 − 17

(159) 67 − 16

(160) 63 − 52

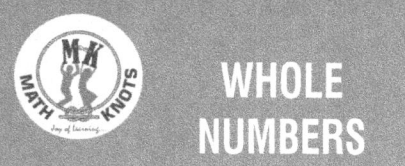

WHOLE NUMBERS

Basic Math

Find the difference of the numbers given below.

(161) 55 − 52

(162) 30 − 18

(163) 62 − 5

(164) 88 − 22

(165) 44 − 44

(166) 97 − 59

(167) 16 − 2

(168) 100 − 31

(169) 56 − 47

(170) 54 − 27

WHOLE NUMBERS

Basic Math

Find the difference of the numbers given below.

(171) 63 − 22

(172) 46 − 8

(173) 84 − 7

(174) 66 − 21

(175) 52 − 14

(176) 89 − 87

(177) 83 − 77

(178) 76 − 38

(179) 89 − 24

(180) 84 − 60

WHOLE NUMBERS

Basic Math

Find the difference of the numbers given below.

(181) 33 − 10

(182) 59 − 4

(183) 71 − 19

(184) 42 − 29

(185) 62 − 17

(186) 47 − 30

(187) 86 − 5

(188) 62 − 42

(189) 29 − 16

(190) 63 − 41

Find the difference of the numbers given below.

(191) 73 − 63

(192) 20 − 16

(193) 99 − 11

(194) 98 − 42

(195) 71 − 18

(196) 37 − 0

(197) 59 − 8

(198) 62 − 22

(199) 13 − 13

(200) 50 − 31

WHOLE NUMBERS

Basic Math

Find the difference of the numbers given below.

(201) 308 − 234

(202) 288 − 153

(203) 393 − 187

(204) 437 − 19

(205) 259 − 21

(206) 366 − 34

(207) 217 − 168

(208) 266 − 120

(209) 362 − 20

(210) 169 − 147

WHOLE NUMBERS

Basic Math

Find the difference of the numbers given below.

(211) 393 − 150

(212) 302 − 34

(213) 234 − 132

(214) 125 − 6

(215) 301 − 124

(216) 317 − 122

(217) 336 − 251

(218) 367 − 1

(219) 368 − 180

(220) 298 − 236

WHOLE NUMBERS

Basic Math

Find the difference of the numbers given below.

(221) 460 − 13

(222) 182 − 141

(223) 217 − 84

(224) 321 − 227

(225) 439 − 352

(226) 298 − 7

(227) 500 − 282

(228) 497 − 132

(229) 415 − 346

(230) 434 − 247

Find the difference of the numbers given below.

(231) 404 − 21

(232) 317 − 84

(233) 462 − 448

(234) 165 − 43

(235) 498 − 147

(236) 324 − 270

(237) 422 − 418

(238) 354 − 97

(239) 342 − 112

(240) 376 − 159

WHOLE NUMBERS

Basic Math

Find the difference of the numbers given below.

(241) 488 − 317   (242) 491 − 400

(243) 360 − 165   (244) 491 − 398

(245) 311 − 56   (246) 454 − 406

(247) 283 − 27   (248) 469 − 359

(249) 416 − 413   (250) 308 − 120

WHOLE NUMBERS

Basic Math

Find the difference of the numbers given below.

(251) 97 − 22 − 52

(252) 86 − 60 − 16

(253) 80 − 5 − 9

(254) 54 − 27 − 22

(255) 90 − 1 − 88

(256) 72 − 19 − 27

(257) 75 − 11 − 0

(258) 75 − 12 − 7

(259) 79 − 1 − 24

(260) 50 − 21 − 25

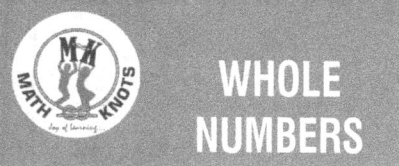

WHOLE NUMBERS

Basic Math

Find the difference of the numbers given below.

(261) 92 − 16 − 56

(262) 90 − 23 − 39

(263) 73 − 2 − 61

(264) 98 − 7 − 37

(265) 71 − 1 − 36

(266) 88 − 50 − 14

(267) 92 − 49 − 35

(268) 85 − 63 − 11

(269) 94 − 57 − 5

(270) 85 − 23 − 53

WHOLE NUMBERS

Basic Math

Find the difference of the numbers given below.

(271) 88 − 1 − 70

(272) 65 − 51 − 0

(273) 95 − 70 − 17

(274) 95 − 19 − 0

(275) 91 − 24 − 29

(276) 93 − 9 − 54

(277) 78 − 25 − 53

(278) 88 − 16 − 71

(279) 87 − 14 − 23

(280) 66 − 53 − 3

WHOLE NUMBERS

Basic Math

Find the difference of the numbers given below.

(281) 78 − 1 − 19

(282) 37 − 2 − 21

(283) 90 − 5 − 53

(284) 91 − 8 − 18

(285) 41 − 27 − 5

(286) 79 − 36 − 6

(287) 87 − 54 − 6

(288) 99 − 8 − 54

(289) 97 − 6 − 59

(290) 87 − 16 − 8

WHOLE NUMBERS

Basic Math

Find the difference of the numbers given below.

(291) 96 − 8 − 35

(292) 62 − 0 − 26

(293) 99 − 47 − 26

(294) 74 − 42 − 19

(295) 69 − 25 − 35

(296) 85 − 37 − 4

(297) 74 − 57 − 12

(298) 75 − 10 − 0

(299) 43 − 0 − 7

(300) 25 − 5 − 12

WHOLE NUMBERS

Basic Math

Evaluate the numerical expressions given below.

(301) 827 + 270 − 798

(302) 104 + 840 + 926

(303) 425 − 381 + 629

(304) 395 + 394 + 719

(305) 874 − 700 + 468

(306) 935 − 819 + 382

(307) 984 + 695 − 380

(308) 178 + 892 − 337

(309) 806 + 116 − 385

(310) 838 − 471 + 148

WHOLE NUMBERS

Basic Math

Evaluate the numerical expressions given below.

(311) $793 - 712 + 506$

(312) $596 + 10 + 743$

(313) $802 + 482 + 231$

(314) $839 - 498 - 321$

(315) $943 + 128 - 979$

(316) $718 + 697 + 58$

(317) $328 - 134 + 484$

(318) $374 - 104 - 22$

(319) $546 - 259 + 362$

(320) $854 + 605 + 29$

WHOLE NUMBERS

Evaluate the numerical expressions given below.

(321) 870 − 173 − 500

(322) 158 + 939 + 449

(323) 999 − 37 + 69

(324) 670 − 521 + 475

(325) 621 − 284 + 141

(326) 179 + 507 − 441

(327) 778 − 123 − 414

(328) 261 + 809 + 856

(329) 628 + 179 − 414

(330) 681 − 21 + 586

Evaluate the numerical expressions given below.

(331) 333 + 819 − 79

(332) 211 + 613 + 38

(333) 735 + 438 − 380

(334) 938 + 554 + 182

(335) 147 + 353 + 908

(336) 352 + 308 − 185

(337) 903 + 419 − 895

(338) 856 − 60 − 550

(339) 463 − 94 + 962

(340) 194 + 857 + 974

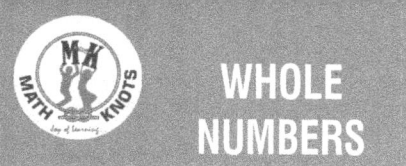

WHOLE NUMBERS

Basic Math

Evaluate the numerical expressions given below.

(341) 451 + 353 + 380

(342) 950 − 309 + 113

(343) 510 + 997 − 130

(344) 968 + 814 + 460

(345) 779 + 763 + 991

(346) 35 + 977 + 549

(347) 415 − 38 − 294

(348) 735 + 104 + 605

(349) 259 − 59 + 83

(350) 360 + 253 + 779

WHOLE NUMBERS

Basic Math

Evaluate the numerical expressions given below.

(351)   857 + 843 + 255

(352)   79 + 152 + 374

(353)   330 + 621 – 939

(354)   166 + 932 + 3

(355)   946 + 631 – 679

(356)   510 + 118 + 142

(357)   300 – 47 – 50

(358)   608 + 777 – 505

(359)   853 – 754 + 758

(360)   834 – 34 + 737

WHOLE NUMBERS

Evaluate the numerical expressions given below.

(361) 556 + 210 + 626

(362) 917 + 173 − 214

(363) 832 − 132 − 152

(364) 918 + 798 + 216

(365) 356 + 562 + 283

(366) 962 + 39 − 753

(367) 509 + 330 − 303

(368) 910 + 763 + 616

(369) 100 + 229 − 173

(370) 283 + 974 + 687

WHOLE NUMBERS

Basic Math

Evaluate the numerical expressions given below.

(371) 554 + 276 + 616

(372) 577 − 289 + 316

(373) 729 + 474 − 0

(374) 879 + 384 + 121

(375) 937 − 161 − 124

(376) 724 + 756 + 743

(377) 972 + 240 − 262

(378) 764 + 466 − 219

(379) 498 + 127 + 179

(380) 283 + 66 + 650

WHOLE NUMBERS

Evaluate the numerical expressions given below.

(381) 805 − 462 − 9

(382) 572 + 629 − 993

(383) 460 + 117 + 32

(384) 860 − 616 − 121

(385) 769 − 696 + 257

(386) 303 + 559 + 925

(387) 353 + 99 + 51

(388) 547 + 37 + 378

(389) 590 − 75 + 79

(390) 833 − 728 + 978

WHOLE NUMBERS

Basic Math

Find the product of the numbers given below.

(391) 35 × 25

(392) 24 × 3

(393) 35 × 21

(394) 33 × 30

(395) 9 × 31

(396) 30 × 17

(397) 31 × 14

(398) 16 × 6

(399) 30 × 5

(400) 15 × 24

WHOLE NUMBERS

Basic Math

Find the product of the numbers given below.

(401) 4 × 6

(402) 26 × 18

(403) 8 × 19

(404) 22 × 6

(405) 14 × 18

(406) 17 × 32

(407) 6 × 19

(408) 24 × 21

(409) 28 × 8

(410) 14 × 23

WHOLE NUMBERS

Basic Math

Find the product of the numbers given below.

(411) 18 × 13

(412) 28 × 0

(413) 13 × 15

(414) 35 × 31

(415) 4 × 32

(416) 9 × 7

(417) 32 × 30

(418) 16 × 12

(419) 15 × 17

(420) 14 × 35

Basic Math

WHOLE NUMBERS

Find the product of the numbers given below.

(421) 17 × 33

(422) 7 × 18

(423) 24 × 7

(424) 17 × 29

(425) 33 × 25

(426) 22 × 16

(427) 16 × 30

(428) 26 × 35

(429) 35 × 7

(430) 17 × 12

WHOLE NUMBERS

Find the product of the numbers given below.

(431) 19 × 31

(432) 20 × 11

(433) 30 × 13

(434) 33 × 33

(435) 12 × 11

(436) 5 × 33

(437) 12 × 35

(438) 13 × 30

(439) 24 × 32

(440) 9 × 28

Find the product of the numbers given below.

(441) 17 × 0

(442) 48 × 21

(443) 41 × 32

(444) 31 × 48

(445) 39 × 38

(446) 43 × 23

(447) 9 × 31

(448) 6 × 28

(449) 29 × 2

(450) 31 × 15

WHOLE NUMBERS

Basic Math

Find the product of the numbers given below.

(451) 15 × 5

(452) 5 × 44

(453) 21 × 49

(454) 11 × 27

(455) 13 × 12

(456) 33 × 24

(457) 22 × 43

(458) 43 × 34

(459) 3 × 35

(460) 24 × 26

WHOLE NUMBERS

Basic Math

Find the product of the numbers given below.

(461) 38 × 26

(462) 26 × 45

(463) 7 × 8

(464) 16 × 44

(465) 33 × 50

(466) 21 × 3

(467) 19 × 4

(468) 10 × 25

(469) 41 × 19

(470) 6 × 50

WHOLE NUMBERS

Basic Math

Find the product of the numbers given below.

(471)   46 × 21

(472)   20 × 40

(473)   12 × 21

(474)   19 × 8

(475)   34 × 15

(476)   41 × 48

(477)   3 × 6

(478)   50 × 2

(479)   34 × 8

(480)   49 × 16

WHOLE NUMBERS

Find the product of the numbers given below.

(481) 13 × 23

(482) 8 × 11

(483) 50 × 47

(484) 48 × 29

(485) 48 × 32

(486) 7 × 34

(487) 15 × 9

(488) 31 × 46

(489) 31 × 12

(490) 35 × 11

WHOLE NUMBERS

Basic Math

Find the product of the numbers given below.

(491) 13 × 14 × 7

(492) 0 × 6 × 19

(493) 11 × 6 × 2

(494) 20 × 5 × 19

(495) 7 × 20 × 2

(496) 15 × 14 × 8

(497) 2 × 18 × 8

(498) 10 × 15 × 0

(499) 10 × 2 × 5

(500) 5 × 11 × 2

Find the product of the numbers given below.

(501) 7 × 0 × 10

(502) 0 × 13 × 3

(503) 7 × 5 × 14

(504) 16 × 8 × 5

(505) 15 × 11 × 8

(506) 12 × 0 × 10

(507) 14 × 11 × 14

(508) 15 × 11 × 4

(509) 7 × 7 × 2

(510) 16 × 19 × 13

Basic Math

WHOLE NUMBERS

Find the product of the numbers given below.

(511) 16 × 14 × 2

(512) 9 × 14 × 13

(513) 0 × 20 × 15

(514) 17 × 15 × 16

(515) 0 × 8 × 16

(516) 18 × 7 × 17

(517) 19 × 4 × 17

(518) 4 × 2 × 10

(519) 9 × 7 × 15

(520) 5 × 7 × 3

WHOLE NUMBERS

Basic Math

Find the product of the numbers given below.

(521) 18 × 14 × 3

(522) 7 × 19 × 18

(523) 7 × 6 × 15

(524) 4 × 14 × 5

(525) 8 × 4 × 17

(526) 14 × 10 × 6

(527) 7 × 20 × 11

(528) 4 × 7 × 15

(529) 12 × 11 × 0

(530) 2 × 6 × 8

WHOLE NUMBERS

Basic Math

Find the product of the numbers given below.

(531) 4 × 6 × 11

(532) 4 × 15 × 10

(533) 8 × 2 × 10

(534) 7 × 12 × 18

(535) 0 × 10 × 11

(536) 14 × 2 × 2

(537) 19 × 10 × 5

(538) 16 × 18 × 19

(539) 14 × 10 × 14

(540) 3 × 16 × 14

WHOLE NUMBERS

Basic Math

Find the quotient of the numbers given below.

(541) 66 ÷ 11

(542) 0 ÷ 13

(543) 0 ÷ 5

(544) 60 ÷ 10

(545) 91 ÷ 7

(546) 60 ÷ 12

(547) 60 ÷ 4

(548) 180 ÷ 12

(549) 63 ÷ 7

(550) 24 ÷ 3

WHOLE NUMBERS

Find the quotient of the numbers given below.

(551) 27 ÷ 9

(552) 70 ÷ 14

(553) 104 ÷ 8

(554) 8 ÷ 4

(555) 96 ÷ 8

(556) 130 ÷ 13

(557) 143 ÷ 13

(558) 112 ÷ 14

(559) 195 ÷ 13

(560) 156 ÷ 12

WHOLE NUMBERS

Basic Math

Find the quotient of the numbers given below.

(561) 98 ÷ 7

(562) 36 ÷ 9

(563) 12 ÷ 6

(564) 120 ÷ 8

(565) 80 ÷ 8

(566) 24 ÷ 2

(567) 36 ÷ 12

(568) 98 ÷ 14

(569) 10 ÷ 5

(570) 39 ÷ 3

WHOLE NUMBERS

Find the quotient of the numbers given below.

(571) 40 ÷ 4

(572) 56 ÷ 7

(573) 54 ÷ 9

(574) 90 ÷ 6

(575) 16 ÷ 4

(576) 35 ÷ 7

(577) 72 ÷ 9

(578) 0 ÷ 4

(579) 12 ÷ 2

(580) 9 ÷ 3

WHOLE NUMBERS

Basic Math

Find the quotient of the numbers given below.

(581)   24 ÷ 12

(582)   168 ÷ 12

(583)   110 ÷ 10

(584)   26 ÷ 2

(585)   0 ÷ 15

(586)   33 ÷ 11

(587)   42 ÷ 14

(588)   0 ÷ 3

(589)   44 ÷ 11

(590)   225 ÷ 15

Find the quotient of the numbers given below.

(591) 267 ÷ 89

(592) 7644 ÷ 98

(593) 2856 ÷ 84

(594) 798 ÷ 19

(595) 980 ÷ 28

(596) 128 ÷ 32

(597) 1470 ÷ 30

(598) 897 ÷ 23

(599) 150 ÷ 30

(600) 9408 ÷ 98

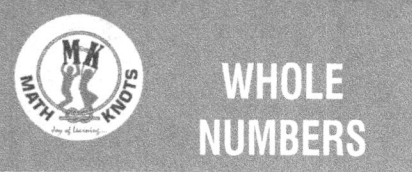

WHOLE NUMBERS

Basic Math

Find the quotient of the numbers given below.

601) 1200 ÷ 100

(602) 2212 ÷ 79

(603) 4655 ÷ 95

(604) 1066 ÷ 13

(605) 630 ÷ 63

(606) 630 ÷ 42

(607) 260 ÷ 13

(608) 5440 ÷ 85

(609) 8316 ÷ 99

(610) 30 ÷ 6

WHOLE NUMBERS

Basic Math

Find the quotient of the numbers given below.

(611)   1392 ÷ 29

(612)   1591 ÷ 43

(613)   333 ÷ 9

(614)   4368 ÷ 48

(615)   1128 ÷ 24

(616)   426 ÷ 6

(617)   54 ÷ 2

(618)   2706 ÷ 82

(619)   2552 ÷ 44

(620)   4611 ÷ 87

WHOLE NUMBERS

Basic Math

Find the quotient of the numbers given below.

(621) 1092 ÷ 12   (622) 728 ÷ 8

(623) 4940 ÷ 52   (624) 2847 ÷ 73

(625) 1235 ÷ 19   (626) 3960 ÷ 72

(627) 2622 ÷ 57   (628) 4900 ÷ 50

(629) 4838 ÷ 82   (630) 546 ÷ 42

WHOLE NUMBERS

Find the quotient of the numbers given below.

(631) 2576 ÷ 28

(632) 1089 ÷ 33

(633) 2716 ÷ 97

(634) 949 ÷ 73

(635) 1860 ÷ 30

(636) 315 ÷ 35

(637) 4200 ÷ 75

(638) 2808 ÷ 72

(639) 605 ÷ 55

(640) 2295 ÷ 51

WHOLE NUMBERS

Basic Math

Round each number given below to the place underlined.

(641)   4<u>4</u>,432

(642)   142,<u>6</u>81.4

(643)   3<u>6</u>9,019

(644)   9<u>1</u>.5

(645)   <u>9</u>9,714.3

(646)   428,75<u>9</u>.84

(647)   445,0<u>9</u>3.123

(648)   <u>9</u>3,698

(649)   215,57<u>9</u>.72

(650)   <u>9</u>,399

WHOLE NUMBERS

Round each number given below to the place underlined.

(651) 1̲0

(652) 1̲7,960

(653) 32,3̲87

(654) 4,99̲9,450

(655) 375,87̲0.78

(656) 9,4̲66

(657) 2̲7,431

(658) 336̲,541

(659) 68,39̲5

(660) 77,689̲.203

WHOLE NUMBERS

Round each number given below to the place underlined.

(661) 28,3<u>9</u>6,426

(662) 3,<u>6</u>05.7

(663) 86,<u>4</u>00

(664) 61,4<u>9</u>6,814.6

(665) 97<u>0</u>,279

(666) 4,3<u>5</u>2,981

(667) <u>1</u>,934,137

(668) <u>2</u>50,329

(669) <u>9</u>,980,832

(670) <u>9</u>41,258

WHOLE NUMBERS

Round each number given below to the place underlined.

(671) 3<u>5</u>1,648

(672) <u>9</u>0,460

(673) 38,<u>6</u>07,380

(674) 1<u>1</u>5,621

(675) 41,7<u>7</u>0,950

(676) 567,9<u>1</u>2,204

(677) 9<u>9</u>1,922

(678) 1,<u>1</u>02,389

(679) <u>4</u>40,738

(680) 8<u>3</u>,385,567

WHOLE NUMBERS

Basic Math

Round each number given below to the place underlined.

(681) 9<u>9</u>1,774

(682) 211,5<u>6</u>7,223

(683) <u>5</u>73,863

(684) <u>4</u>,738,634

(685) <u>5</u>82,315

(686) <u>1</u>,987,480

(687) 848,6<u>7</u>0,970

(688) 7<u>9</u>7,145

(689) 976,<u>9</u>78,350

(690) 57<u>9</u>,893,754

WHOLE NUMBERS

Round each number given below to the place underlined.

(691) 3,993,624,511

(692) 91,299,777

(693) 7,493,402,013

(694) 9,934,879

(695) 993,599,664

(696) 493,036,807

(697) 9,413,579,786

(698) 769,981,833

(699) 822,812,199

(700) 9,936,995,293

Round each number given below to the place underlined.

(701) 3<u>8</u>,473,746

(702) 3<u>3</u>,316,806

(703) 38<u>0</u>,452,753

(704) 1<u>5</u>8,258,433

(705) 5,7<u>9</u>4,579,737

(706) <u>9</u>07,809,252

(707) 2,22<u>2</u>,260,127

(708) <u>2</u>2,601,278

(709) 9<u>9</u>2,350,490

(710) <u>8</u>6,684,777

WHOLE NUMBERS

Basic Math

Round each number given below to the place underlined.

(711) 4,2<u>8</u>0,332,597

(712) 3,<u>9</u>85,423,746

(713) <u>9</u>67,408,936

(714) <u>9</u>03,431,127

(715) <u>7</u>98,087,350

(716) <u>3</u>,909,184,171

(717) 524,<u>0</u>68,035

(718) <u>7</u>,830,509

(719) 392,9<u>3</u>9,237

(720) 3,7<u>8</u>2,665

WHOLE NUMBERS

Round each number given below to the place underlined.

(721) <u>9</u>,962,376

(722) 5,0<u>3</u>7,826

(723) 38<u>4</u>,805,115

(724) 26<u>1</u>,494,323

(725) 4<u>9</u>,576

(726) 10,<u>5</u>59,308

(727) <u>1</u>,685,741,583

(728) <u>9</u>41,733,706

(729) 6,5<u>9</u>9,245

(730) 74,5<u>0</u>7,219

Round each number given below to the place underlined.

(731) 6,9̲82,307,973

(732) 4̲1,912,041

(733) 9̲92,323,975

(734) 2,403,9̲13

(735) 75̲5,558

(736) 172̲,568,487

(737) 33̲7,569,767

(738) 8,56̲5,146,444

(739) 5,49̲2,491

(740) 248̲,313

WHOLE NUMBERS

Write each number given below in words.

(741) 938

(742) 632

(743) 858

(744) 566

(745) 915

(746) 493

(747) 154

(748) 625

(749) 863

(750) 520

WHOLE NUMBERS

Write each number given below in words.

(751)   295

(752)   991

(753)   377

(754)   126

(755)   889

(756)   773

(757)   584

(758)   941

(759)   153

(760)   859

WHOLE NUMBERS

Basic Math

Write each number given below in words.

(761) 571

(762) 432

(763) 655

(764) 971

(765) 162

(766) 300,566

(767) 500,000

(768) 460,637

(769) 7,013

(770) 590,714

WHOLE NUMBERS

Write each number given below in words.

(771) 980,935

(772) 40,234

(773) 214,203

(774) 647,269

(775) 881,852

(776) 93,272

(777) 25,450

(778) 9,288

(779) 181,329

(780) 964,996

WHOLE NUMBERS

Write each number given below in words.

(781) 2,980

(782) 3,064

(783) 400,507

(784) 74,551

(785) 8,482

(786) 641,731

(787) 106,310

(788) 7,877

(789) 61,655

(790) 6,928

WHOLE NUMBERS

Write each number given below in words.

(791)   666,071,006

(792)   23,004,010

(793)   4,870,001

(794)   5,008,345

(795)   353,403,690

(796)   80,600,800

(797)   30,300,401

(798)   3,000,020

(799)   557,312,000

(800)   330,000,000

WHOLE NUMBERS

Basic Math

Write each number given below in words.

(801) 1,058,660

(802) 36,300,393

(803) 44,090,600

(804) 715,000,025

(805) 5,089,300

(806) 869,140,759

(807) 88,020,084

(808) 11,600,961

(809) 5,890,100

(810) 60,094,960

WHOLE NUMBERS

Write each number given below in words.

(811) 60,050,010

(812) 7,000,006

(813) 41,770,030

(814) 7,409,900

(815) 703,000,715

(816) 70,002,070,830

(817) 4,560,371,062

(818) 500,720,330,041

(819) 38,100,972,145

(820) 73,005,700,500

WHOLE NUMBERS

Basic Math

Write each number given below in words.

(821) 676,510,608,720

(822) 5,165,650,802

(823) 2,006,060,340

(824) 600,003,030,020

(825) 40,000,075,000

(826) 4,005,600,106

(827) 800,058,908,400

(828) 250,010,406,048

(829) 900,595,050,081

(830) 1,409,936,600

Write each number given below in words.

(831) 90,030,010,010

(832) 64,100,230,000

(833) 10,607,806,106

(834) 14,750,305,080

(835) 601,800,518,005

(836) 1,020,040,918

(837) 1,300,407,830

(838) 709,549,000,733

(839) 53,101,705,470

(840) 9,094,080,009

Write each number given below in words.

(841) 630,016,040,300,000

(842) 4,070,330,093,000

(843) 1,644,008,004,138

(844) 75,720,550,900,754

(845) 354,300,102,001,550

(846) 1,300,102,001,550

(847) 30,009,071,300,700

(848) 404,200,960,073,050

(849) 190,700,080,880,253

(850) 13,495,005,000,340

WHOLE NUMBERS

Basic Math

Write each number given below in words.

(851)   2,400,977,100,036

852)   900,203,001,000,400

(853)   70,002,009,500,202

(854)   80,310,679,170,505

(855)   92,840,904,001,400

(856)   700,007,730,004,032

(857)   7,505,172,000,000

(858)   1,382,003,087,050

(859)   4,900,010,200,155

(860)   5,020,180,032,106

Write each number given below in words.

861)   81,572,007,002,150

(862)   572,000,454,400,190

(863)   6,660,702,181,786

(864)   190,005,000,984,160

(865)   356,046,700,650,030

WHOLE NUMBERS

Basic Math

Write the words given below as a number.

(866) seven hundred sixty-three

(867) nine hundred fifty-two

(868) nine hundred one

(869) nine hundred ninety-three

(870) four hundred ninety-one

(871) two hundred fifty-six

(872) four hundred five

(873) six hundred ninety-four

(874) nine hundred eighty-seven

(875) four hundred thirty

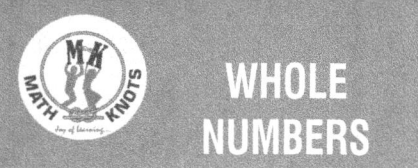

WHOLE NUMBERS

Basic Math

Write the words given below as a number.

(876) eight hundred seventeen

(877) five hundred twenty-six

(878) six hundred eight

(879) six hundred forty-one

(880) two hundred forty-four

(881) three hundred three

(882) nine hundred eighty-nine

(883) three hundred fifty-seven

(884) eight hundred forty-eight

(885) five hundred sixty

WHOLE NUMBERS

Basic Math

Write the words given below as a number.

(886) eight hundred six

(887) two hundred ninety-seven

(888) two hundred fifty-eight

(889) one hundred seventy-two

(890) four hundred sixty-three

(891) nine hundred nine thousand, ninety

(892) five thousand, two hundred seven

(893) nine thousand, five hundred six

(894) seventy thousand, six

(895) nine hundred forty-seven thousand, seven hundred one

WHOLE NUMBERS

Basic Math

Write the words given below as a number.

(896) eight hundred thousand, thirty

(897) twenty-one thousand, ninety

(898) six hundred thousand, seven hundred

(899) two hundred six thousand, three hundred forty

(900) seven thousand, six hundred

(901) eighty thousand, six hundred

(902) five hundred thousand, seven hundred ninety

(903) three hundred forty-three thousand, eight hundred fifty-four

(904) nine thousand, six hundred

WHOLE NUMBERS

Basic Math

Write the words given below as a number.

(905) six thousand, one hundred twenty

(906) eight thousand, seven hundred

(907) twenty thousand, one hundred nineteen

(908) seventy-nine thousand, thirty

(909) seventeen thousand, eight hundred

(910) six hundred fifty thousand, one hundred sixty-five

(911) three thousand, four hundred

(912) ninety thousand, seven hundred sixty-four

(913) fifty-three thousand, five hundred

(914) ten thousand, sixty

WHOLE NUMBERS

Basic Math

Write the words given below as a number.

(915) ninety-two thousand, two hundred

(916) forty-eight million, one hundred eighty-seven thousand, two

(917) three hundred three million, eight hundred thousand

(918) five hundred sixty million, thirty thousand, nine

(919) seven million, thirty-five thousand, nine hundred

WHOLE NUMBERS

Write the words given below as a number.

(920) six million, nine hundred thousand, seven hundred nine

(921) one million, forty-five thousand, nine hundred ninety-eight

(922) six million, one thousand, ninety

(923) two million, two hundred six thousand, five

(924) three million, ninety-six thousand, six

Write the words given below as a number.

(925) one million, four hundred

(926) six million, eight hundred twenty-eight thousand, six

(927) ten million, five hundred fifteen thousand, three hundred

(928) sixty-nine million, five hundred thirty-one thousand

(929) seventy-four million, nine hundred thousand, ninety

WHOLE NUMBERS

Basic Math

Write the words given below as a number.

(930) two hundred sixty million, seven hundred two thousand, twenty-six

(931) forty-four million, four hundred sixty-two thousand, forty-eight

(932) two million, eight hundred sixty-two thousand, six

(933) one million, ninety thousand, four

(934) nine million, one hundred seven thousand, six hundred

WHOLE NUMBERS

Basic Math

Write the words given below as a number.

(935)   ninety-seven million, five hundred seventy-two thousand, six hundred six

(936)   one hundred thirty-one million, seven hundred nine thousand, six hundred three

(937)   four million, ninety thousand, two

(938)   three hundred nine million, one hundred twenty thousand, ninety

(939)   five million, three hundred eight thousand, one hundred nine

WHOLE NUMBERS

Basic Math

Write the words given below as a number.

(940)   eighty-five million, four hundred thousand, forty

(941)   seventy billion, ninety thousand, one hundred ninety-four

(942)   four hundred ninety-one billion, eight hundred sixty-two million

(943)   twenty-eight billion, four hundred ten million, five hundred eighty-nine thousand, six hundred eighty

(944)   three billion, six hundred million, sixty-six thousand, five hundred ten

WHOLE NUMBERS — Basic Math

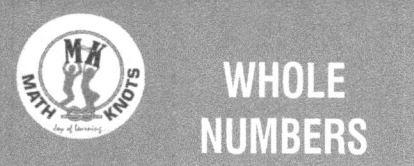

Write the words given below as a number.

(945) eighty-one billion, two thousand, one hundred thirty-four

(946) nine hundred sixty-three billion, eight hundred fifty million, four hundred three thousand, one

(947) thirty-six billion, eight million, six hundred one thousand, fifty-six

(948) twenty billion, six million, six hundred thousand, twelve

(949) seventy billion, eight hundred seventy million, two hundred thousand, eight hundred eighty-one

WHOLE NUMBERS

Basic Math

Write the words given below as a number.

(950) two billion, one hundred thirty-four million, nine thousand, eight hundred sixty-nine

(951) five hundred eighty billion, seven hundred ninety-five million, five hundred twenty-three thousand, nineteen

(952) one hundred six billion, one hundred twenty-three million, three hundred six thousand, eight hundred five

(953) seventy-eight billion, nine hundred thousand, seventy

(954) seven hundred billion, three hundred million, seven hundred five thousand, thirty-three

WHOLE NUMBERS

Write the words given below as a number.

(955) seven billion, two hundred million, eight hundred sixty-three thousand, sixty-three

(956) forty-three billion, sixty million, three hundred ninety-eight thousand, eight hundred ninety

(957) two billion, one hundred forty-one million, two hundred thousand, five hundred nine

(958) sixty-nine billion, ninety-eight million, seven hundred seven thousand, four hundred three

(959) forty billion, five hundred eighty million, nine hundred six thousand

WHOLE NUMBERS

Basic Math

Write the words given below as a number.

(960) nine billion, five hundred eight thousand, three hundred ninety-one

(961) forty-one billion, one hundred sixty-one million, eight hundred eighty thousand, seven hundred sixty-eight

(962) three billion, four million, one hundred twenty-three thousand, three hundred forty-six

(963) eighty billion, one hundred three million, five hundred thousand, seven hundred ninety-five

(964) seven hundred thirteen billion, two hundred thirty million, seven hundred thousand, eight

WHOLE NUMBERS

Write the words given below as a number.

(965)   eight billion, ninety-eight million, seven hundred

(966)   five hundred thirty trillion, ninety billion, nine hundred nineteen million, five thousand

(967)   eighty-eight trillion, six hundred fifty-nine billion, six hundred two million, five hundred forty thousand, twenty

(968)   four trillion, ninety billion, sixty-two million, six hundred sixty-six thousand

(969)   eight hundred two trillion, seven hundred seventy billion, two hundred eight million, seven hundred seventy-five thousand

WHOLE NUMBERS

Write the words given below as a number.

(970) six trillion, sixty billion, seven hundred forty-four million, five hundred eighty thousand, four hundred thirty-two

(971) seventy trillion, eight billion, one hundred eighty million, eight hundred seventy thousand, one

(972) eleven trillion, seven hundred sixteen billion, one hundred five million, two thousand, eight

(973) twenty trillion, one hundred ninety-four billion, eight million, five hundred thirty thousand, sixty

(974) sixteen trillion, thirty-four billion, five million, seven hundred six thousand, three hundred seventy

WHOLE NUMBERS

Write the words given below as a number.

(975) twenty-five trillion, two hundred billion, one million, four hundred forty thousand, four hundred

(976) eight hundred sixty-nine trillion, twenty-one billion, twenty million, nine hundred thirty-two thousand, five

(977) six trillion, three hundred one billion, seventy-six million, one hundred seven thousand, two hundred

(978) ten trillion, two billion, three hundred sixty million, seven hundred ten thousand, five hundred two

(979) eighty-three trillion, forty-two billion, four hundred million, six hundred thousand, three hundred sixty

WHOLE NUMBERS

Basic Math

Write the words given below as a number.

(980)   forty trillion, three hundred fifty billion, sixty-four million, seven hundred twelve

(981)   one hundred ninety trillion, two billion, seven hundred eighty million, ninety-one thousand, seven hundred ninety

(982)   nine trillion, three hundred twenty billion, two million, seven hundred eighty thousand, eighty-nine

(983)   ten trillion, eighty million, four thousand, thirty

(984)   six trillion, ten billion, seventy million, nine hundred thirty-three thousand, five hundred five

WHOLE NUMBERS

Basic Math

Write the words given below as a number.

(985) five hundred trillion, five hundred ninety billion, three hundred seventy-five million, sixty-five thousand, thirty-two

(986) twenty trillion, thirty-seven billion, eighty-three million, six hundred one thousand, two hundred

(987) one trillion, thirty-five billion, three hundred million, four hundred thirty thousand

(988) ninety-seven trillion, six hundred forty billion, five hundred twelve million, eight hundred

(989) forty trillion, six hundred seven billion, ninety-five million, forty thousand

Write the words given below as a number.

(990) three trillion, four billion, seven thousand, seven hundred one

WHOLE NUMBERS

Basic Math

Write the name of each number place as underlined.

(991)   4,10<u>1</u>,606

(992)   <u>5</u>6,001

(993)   <u>3</u>59

(994)   765,6<u>1</u>4,933

(995)   93,08<u>5</u>,437

(996)   3,<u>7</u>48

(997)   119,0<u>3</u>0,008

(998)   50,4<u>6</u>1,971

(999)   35<u>6</u>,219

(1000) 6,2<u>2</u>0,820

WHOLE NUMBERS

Basic Math

Write the name of each number place as underlined.

(1001)   42,3<u>9</u>2,597

(1002)   18,46<u>9</u>,975

(1003)   16,115,<u>6</u>85

(1004)   7,69<u>8</u>,692

(1005)   <u>1</u>,476

(1006)   6<u>2</u>,138

(1007)   670,4<u>3</u>3,109

(1008)   <u>4</u>93

(1009)   <u>9</u>2,970

(1010)   <u>8</u>7,821

WHOLE NUMBERS

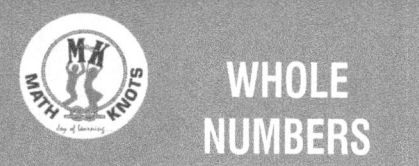

Basic Math

Write the name of each number place as underlined.

(1011) 24,768,323

(1012) 396,718

(1013) 937,269

(1014) 97,658

(1015) 39,521

(1016) 8,356,219

(1017) 99,464

(1018) 581,594

(1019) 484,130

(1020) 452,375

WHOLE NUMBERS

Write the name of each number place as underlined.

(1021) 3̲ 4,800

(1022) 9̲,872

(1023) 663,4̲93

(1024) 715,5̲43

(1025) 42 8̲,523

(1026) 9, 8̲56

(1027) 964,6̲60

(1028) 79,62̲6,511

(1029) 359,65̲8,719

(1030) 1̲45

WHOLE NUMBERS

Basic Math

Write the name of each number place as underlined.

(1031)   845,8̲93

(1032)   6̲64

(1033)   1̲,258

(1034)   7̲,732

(1035)   3,80̲3,742

(1036)   2̲63

(1037)   581̲,534

(1038)   590̲,351

(1039)   3,902̲,131

(1040)   95̲1,432

WHOLE NUMBERS

Basic Math

Write the name of each number place as underlined.

(1041) 7<u>5</u>1,416,665

(1042) <u>4</u>6,488,410

(1043) 5<u>7</u>4,216,722

(1044) <u>4</u>,783,220

(1045) 4<u>1</u>,558,199

(1046) <u>8</u>,756,149

(1047) 9,8<u>3</u>5,175,882

(1048) 3,4<u>1</u>0,453,874

(1049) 64,<u>0</u>48,130

(1050) <u>9</u>5,905,980

WHOLE NUMBERS

Basic Math

Write the name of each number place as underlined.

(1051) 5̲5,814,624

(1052) 95,5̲74,873

(1053) 17 0̲,220,639

(1054) 4̲,633,671

(1055) 5̲3,105,317

(1056) 9̲54,176

(1057) 932,8̲80,806

(1058) 69̲,950,349

(1059) 3,34̲3,275,182

(1060) 45,0̲97,312

WHOLE NUMBERS

Basic Math

Write the name of each number place as underlined.

(1061) 80,4̲97,131

(1062) 6̲4,210,021

(1063) 9̲,881,600

(1064) 20,1̲01,616

(1065) 19 3̲,257,555

(1066) 7̲,702,838

(1067) 5̲2,268,766

(1068) 1̲,743,445

(1069) 7, 2̲44,916

(1070) 21 5̲,275,675

WHOLE NUMBERS

Basic Math

Write the name of each number place as underlined.

(1071)   <u>9</u>9,418,445

(1072)   9<u>2</u>9,113,722

(1073)   <u>8</u>49,436

(1074)   <u>8</u> 3,616,175

(1075)   <u>2</u>0,446,223

(1076)   43<u>2</u>,158,219

(1077)   4, <u>0</u>51,302

(1078)   1,3<u>4</u>3,052,394

(1079)   27, <u>8</u>92,574

(1080)   4, <u>0</u>27,031

WHOLE NUMBERS

Basic Math

Write the name of each number place as underlined.

(1081) 70<u>0</u>,016,059

(1082) <u>4</u>,550,202

(1083) 1,1<u>2</u>6,594,037

(1084) <u>1</u>2,308,265

(1085) 2<u>2</u> 3,883,363

(1086) <u>7</u>6,003,086

(1087) 49,<u>5</u>38,335

(1088) 1<u>8</u>2,777,379

(1089) 3,<u>2</u>53,157

(1090) 4<u>1</u>8,498,217

WHOLE NUMBERS

Basic Math

Write the name of each number place as underlined.

(1091)   7,2̲00,401,177

(1092)   7̲30,956,995

(1093)   1̲31,084,972

(1094)   3̲,947,019,727

(1095)   3,4 7̲1,418,431

(1096)   6,6 0̲1,513,299

(1097)   4̲,583,580,744

(1098)   6 4̲6,698,330

(1099)   99̲1,771,708

(1100)   3,1 4̲8,400,883

WHOLE NUMBERS

Write the name of each number place as underlined.

(1101) <u>1</u>,912,854,494

(1102) 81<u>2</u>,267,734

(1103) 9,<u>9</u>69,492,283

(1104) <u>1</u>,821,843,412

(1105) <u>1</u>1,202,808

(1106) <u>5</u>5,397,039

(1107) 1<u>4</u>4,221,446

(1108) <u>5</u>,213,489,141

(1109) 1,<u>1</u>92,487,901

(1110) <u>6</u>,521,208,621

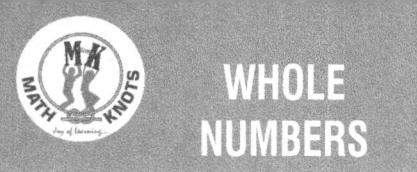

WHOLE NUMBERS

Basic Math

Write the name of each number place as underlined.

(1111)   <u>3</u>,539,861,887

(1112)   <u>3</u>,342,174,736

(1113)   1<u>3</u>4,228,733

(1114)   <u>4</u>,370,601,267

(1115)   9<u>1</u>8,313,646

(1116)   <u>4</u>93,696,863

(1117)   <u>4</u>,565,545,775

(1118)   <u>2</u>,508,126,115

(1119)   <u>3</u>,849,110,869

(1120)   <u>6</u>4,599,323

WHOLE NUMBERS

Write the name of each number place as underlined.

(1121) 4, 4̲51,837,516

(1122) 3̲57,763,995

(1123) 6, 5̲94,485,409

(1124) 3̲5,900,360

(1125) 7̲ ,614,910,769

(1126) 3̲4,080,413

(1127) 9̲2,398,785

(1128) 3̲,338,067,087

(1129) 2̲ ,195,481,825

(1130) 3 2̲6,249,568

WHOLE NUMBERS

Basic Math

Write the name of each number place as underlined.

(1131) 9,3<u>5</u>3,532,175

(1132) 9,<u>0</u>90,518,590

(1133) 1,<u>7</u>38,548,575

(1134) <u>7</u>4,495,941

(1135) 6,<u>2</u>14,397,168

(1136) <u>3</u>46,736,397

(1137) <u>7</u>6,830,166

(1138) 9,<u>9</u>08,581,629

(1139) <u>1</u>8,594,071

(1140) <u>9</u>47,234,267

Round each number given below as mentioned.

(1141) 401,969 ; hundreds

(1142) 12,547,759 ; thousands

(1143) 49,916 ; thousands

(1144) 321,337 ; thousands

(1145) 25,437,028; hundreds

(1146) 39,321,240 ; thousands

(1147) 365,532 ; thousands

(1148) 993,775 ; ten thousands

(1149) 782,908 ; thousands

(1150) 187,654 ; ten thousands

WHOLE NUMBERS

Basic Math

Round each number given below as mentioned.

(1151) 7,222 ; hundreds

(1152) 92,492,862 ; ten thousands

(1153) 837,347 ; hundreds

(1154) 272,616 ; thousands

(1155) 81,664,935 ; ten thousands

(1156) 705,837 ; hundreds

(1157) 597,186 ; thousands

(1158) 7,382,201 ; ten thousands

(1159) 34,708 ; thousands

(1160) 878,005,696 ; ten thousands

WHOLE NUMBERS

Round each number given below as mentioned.

(1161)   52,831,689 ; hundreds

(1162)   19,810,931 ; thousands

(1163)   6,290,101 ; ten thousands

(1164)   2,474 ; hundreds

(1165)   8,499,023 ; thousands

(1166)   250,630 ; ten thousands

(1167)   589,706 ; hundreds

(1168)   369,316 ; hundreds

(1169)   56,664 ; thousands

(1170)   620,886 ; hundreds

WHOLE NUMBERS

Basic Math

Round each number given below as mentioned.

(1171)   924 ; hundreds

(1172)   6,374 ; thousands

(1173)   4,722 ; thousands

(1174)   8,324,421 ; ten thousands

(1175)   2,971 ; hundreds

(1176)   8,979 ; thousands

(1177)   25,852,554 ; thousands

(1178)   67,693,956 ; ten thousands

(1179)   197,058 ; hundreds

(1180)   4,746,535 ; ten thousands

WHOLE NUMBERS

Round each number given below as mentioned.

(1181) 725,912 ; hundreds

(1182) 374,706 ; ten thousands

(1183) 7,739,583 ; thousands

(1184) 85,282,809 ; hundreds

(1185) 4,918 ; hundreds

(1186) 9,412 ; thousands

(1187) 34,482 ; ten thousands

(1188) 67,229,696 ; hundreds

(1189) 95,344,375 ; ten thousands

(1190) 405,266,547 ; ten thousands

Round each number given below as mentioned.

(1191) 72.6167797 ; millionths

(1192) 32.785600 ; thousandths

(1193) 30.0328831 ; millionths

(1194) 5,816,587,264 ; billions

(1195) 28.1987697 ; millionths

(1196) 0.96861 ; ten-thousandths

(1197) 97.730735 ; ten-thousandths

(1198) 991,823,901 ; hundred millions

(1199) 6,513,840 ; hundred thousands

(1200) 0.949755 ; ten-thousandths

WHOLE NUMBERS

Round each number given below as mentioned.

(1201) 5.6335571 ; millionths

(1202) 998,991,804 ; hundred millions

(1203) 929,633,041 ; hundred millions

(1204) 549,089,959 ; millions

(1205) 396,144,084 ; ten millions

(1206) 31,451,665 ; millions

(1207) 997,950,274 ; hundred millions

(1208) 1,986,533,117 ; hundred millions

(1209) 2,315,204,826 ; billions

(1210) 707,868 ; hundred thousands

WHOLE NUMBERS

Round each number given below as mentioned.

(1211) 3,428,392,148 ; billions

(1212) 88.10197 ; thousandths

(1213) 6.973255 ; thousandths

(1214) 6.54997 ; ten-thousandths

(1215) 947,400 ; hundred thousands

(1216) 9.6586260 ; millionths

(1217) 212,276,462 ; ten millions

(1218) 17.2635581 ; millionths

(1219) 0.926104 ; thousandths

(1220) 9,333,742 ; millions

WHOLE NUMBERS

Round each number given below as mentioned.

(1221) 6.74767 ; ten-thousandths

(1222) 64,534,706 ; millions

(1223) 1,876,744 ; millions

(1224) 8,547,859,371 ; billions

(1225) 5,166,512 ; millions

(1226) 119,498,350 ; hundred millions

(1227) 686,995,571 ; ten millions

(1228) 527,973,534 ; hundred millions

(1229) 3.28992 ; ten-thousandths

(1230) 98,039,495 ; ten millions

WHOLE NUMBERS

Round each number given below as mentioned.

(1231) 212,346,465 ; hundred millions

(1232) 116,503 ; hundred thousands

(1233) 1,267,811,942 ; hundred millions

(1234) 36.31734 ; ten-thousandths

(1235) 9.91098 ; ten-thousandths

(1236) 34,126,030 ; ten millions

(1237) 248,313,536 ; hundred millions

(1238) 5.986330 ; ten-thousandths

(1239) 2.279389 ; thousandths

(1240) 75,472,490 ; ten millions

WHOLE NUMBERS

Round each number given below as mentioned.

(1241) 2.6993 ; thousandths

(1242) 9.9493096 ; millionths

(1243) 8.229532 ; ten-thousandths

(1244) 9,241,764 ; millions

(1245) 1.94726 ; ten-thousandths

(1246) 418,950 ; hundred thousands

(1247) 52.976694 ; hundred-thousandths

(1248) 21.398169 ; thousandths

(1249) 7.5245046 ; millionths

(1250) 4.5580718 ; millionths

Round each number given below as mentioned.

(1251)   67,460,409 ; hundred thousands

(1252)   19.221393 ; thousandths

(1253)   784,456 ; hundred thousands

(1254)   1,639,211,882 ; billions

(1255)   1,865,080,980 ; billions

(1256)   7.3625205 ; millionths

(1257)   7,476,177,467 ; billions

(1258)   9,844,959,633 ; billions

(1259)   8,916,312,142 ; billions

(1260)   6.4796 ; thousandths

Round each number given below as mentioned.

(1261) 8,983,520,543 ; hundred millions

(1262) 3.5083178 ; millionths

(1263) 6.442496 ; hundred-thousandths

(1264) 629,161,232 ; millions

(1265) 9,632,757,489 ; billions

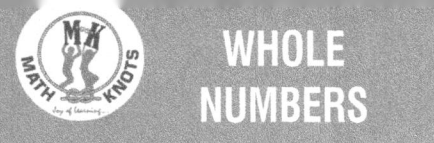

WHOLE NUMBERS

Basic Math Answer Keys

WHOLE NUMBERS

Basic Math Answer Keys

Answer Key

(1) 654   (2) 424   (3) 697   (4) 319

(5) 480   (6) 180   (7) 536   (8) 661

(9) 499   (10) 436   (11) 326   (12) 284

(13) 214   (14) 437   (15) 491   (16) 623

(17) 786   (18) 394   (19) 382   (20) 507

(21) 287   (22) 683   (23) 369   (24) 640

(25) 805   (26) 532   (27) 297   (28) 506

(29) 540   (30) 235   (31) 452   (32) 330

(33) 315   (34) 246   (35) 587   (36) 662

(37) 885   (38) 606   (39) 357   (40) 408

(41) 646   (42) 647   (43) 397   (44) 292

(45) 925   (46) 557   (47) 451   (48) 301

WHOLE NUMBERS

Basic Math Answer Keys

(49) 481   (50) 731   (51) 1290   (52) 1710

(53) 1601   (54) 1403   (55) 1570   (56) 1994

(57) 1841   (58) 1118   (59) 1411   (60) 2164

(61) 1923   (62) 2235   (63) 1663   (64) 1098

(65) 1304   (66) 1763   (67) 1832   (68) 1553

(69) 1514   (70) 986   (71) 1279   (72) 1730

(73) 1337   (74) 860   (75) 1641   (76) 2051

(77) 1707   (78) 626   (79) 1162   (80) 2269

(81) 1525   (82) 2663   (83) 531   (84) 2502

(85) 2787   (86) 1440   (87) 1519   (88) 1426

(89) 2137   (90) 815   (91) 1424   (92) 1272

(93) 1704   (94) 1483   (95) 1893   (96) 1312

WHOLE NUMBERS

Basic Math Answer Keys

(97) 886   (98) 1694   (99) 1247   (100) 869

(101) 1843   (102) 1872   (103) 3140   (104) 2228

(105) 2443   (106) 1464   (107) 2015   (108) 1718

(109) 1370   (110) 2038   (111) 2026   (112) 934

(113) 1970   (114) 2757   (115) 2771   (116) 2001

(117) 2599   (118) 2644   (119) 2128   (120) 2459

(121) 3051   (122) 1890   (123) 2583   (124) 1825

(125) 2119   (126) 2046   (127) 2915   (128) 715

(129) 3185   (130) 1529   (131) 3213   (132) 3257

(133) 2116   (134) 1471   (135) 1067   (136) 2743

(137) 993   (138) 1825   (139) 2115   (140) 2502

(141) 1677   (142) 1119   (143) 2554   (144) 3130

WHOLE NUMBERS

Basic Math Answer Keys

(145) 1498 (146) 1672 (147) 1503 (148) 2118

(149) 2636 (150) 2309 (151) 59   (152) 37

(153) 41   (154) 37   (155) 10   (156) 49

(157) 30   (158) 50   (159) 51   (160) 11

(161) 3   (162) 12   (163) 57   (164) 66

(165) 0   (166) 38   (167) 14   (168) 69

(169) 9   (170) 27   (171) 41   (172) 38

(173) 77   (174) 45   (175) 38   (176) 2

(177) 6   (178) 38   (179) 65   (180) 24

(181) 23   (182) 55   (183) 52   (184) 13

(185) 45   (186) 17   (187) 81   (188) 20

(189) 13   (190) 22   (191) 10   (192) 4

WHOLE NUMBERS

Basic Math Answer Keys

(193) 88   (194) 56   (195) 53   (196) 37

(197) 51   (198) 40   (199) 0   (200) 19

(201) 74   (202) 135   (203) 206   (204) 418

(205) 238   (206) 332   (207) 49   (208) 146

(209) 342   (210) 22   (211) 243   (212) 268

(213) 102   (214) 119   (215) 177   (216) 195

(217) 85   (218) 366   (219) 188   (220) 62

(221) 447   (222) 41   (223) 133   (224) 94

(225) 87   (226) 291   (227) 218   (228) 365

(229) 69   (230) 187   (231) 383   (232) 233

(233) 14   (234) 122   (235) 351   (236) 54

(237) 4   (238) 257   (239) 230   (240) 217

WHOLE NUMBERS

Basic Math Answer Keys

(241) 171   (242) 91   (243) 195   (244) 93

(245) 255   (246) 48   (247) 256   (248) 110

(249) 3   (250) 188   (251) 23   (252) 10

(253) 66   (254) 5   (255) 1   (256) 26

(257) 64   (258) 56   (259) 54   (260) 4

(261) 20   (262) 28   (263) 10   (264) 54

(265) 34   (266) 24   (267) 8   (268) 11

(269) 32   (270) 9   (271) 17   (272) 14

(273) 8   (274) 76   (275) 38   (276) 30

(277) 0   (278) 1   (279) 50   (280) 10

(281) 58   (282) 14   (283) 32   (284) 65

(285) 9   (286) 37   (287) 27   (288) 37

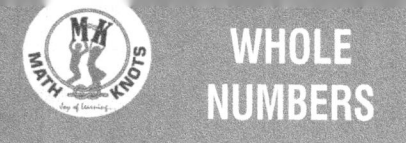

WHOLE NUMBERS

Basic Math Answer Keys

(289) 32   (290) 63   (291) 53   (292) 36

(293) 26   (294) 13   (295) 9   (296) 44

(297) 5   (298) 65   (299) 36   (300) 8

(301) 299   (302) 1870   (303) 673   (304) 1508

(305) 642   (306) 498   (307) 1299   (308) 733

(309) 537   (310) 515   (311) 587   (312) 1349

(313) 1515   (314) 20   (315) 92   (316) 1473

(317) 678   (318) 248   (319) 649   (320) 1488

(321) 197   (322) 1546   (323) 1031   (324) 624

(325) 478   (326) 245   (327) 241   (328) 1926

(329) 393   (330) 1246   (331) 1073   (332) 862

(333) 793   (334) 1674   (335) 1408   (336) 475

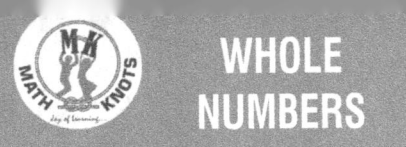

WHOLE NUMBERS

Basic Math Answer Keys

(337) 427   (338) 246   (339) 1331   (340) 2025

(341) 1184   (342) 754   (343) 1377   (344) 2242

(345) 2533   (346) 1561   (347) 83   (348) 1444

(349) 283   (350) 1392   (351) 1955   (352) 605

(353) 12   (354) 1101   (355) 898   (356) 770

(357) 203   (358) 880   (359) 857   (360) 1537

(361) 1392   (362) 876   (363) 548   (364) 1932

(365) 1201   (366) 248   (367) 536   (368) 2289

(369) 156   (370) 1944   (371) 1446   (372) 604

(373) 1203   (374) 1384   (375) 652   (376) 2223

(377) 950   (378) 1011   (379) 804   (380) 999

(381) 334   (382) 208   (383) 609   (384) 123

WHOLE NUMBERS

Basic Math Answer Keys

(385) 330   (386) 1787   (387) 503   (388) 962

(389) 594   (390) 1083   (391) 875   (392) 72

(393) 735   (394) 990   (395) 279   (396) 510

(397) 434   (398) 96   (399) 150   (400) 360

(401) 24   (402) 468   (403) 152   (404) 132

(405) 252   (406) 544   (407) 114   (408) 504

(409) 224   (410) 322   (411) 234   (412) 0

(413) 195   (414) 1085   (415) 128   (416) 63

(417) 960   (418) 192   (419) 255   (420) 490

(421) 561   (422) 126   (423) 168   (424) 493

(425) 825   (426) 352   (427) 480   (428) 910

(429) 245   (430) 204   (431) 589   (432) 220

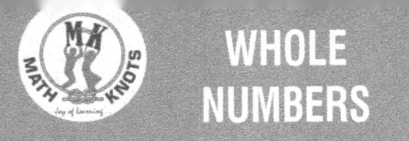

WHOLE NUMBERS

Basic Math Answer Keys

(433) 390   (434) 1089   (435) 132   (436) 165

(437) 420   (438) 390   (439) 768   (440) 252

(441) 0   (442) 1008   (443) 1312   (444) 1488

(445) 1482   (446) 989   (447) 279   (448) 168

(449) 58   (450) 465   (451) 75   (452) 220

(453) 1029   (454) 297   (455) 156   (456) 792

(457) 946   (458) 1462   (459) 105   (460) 624

(461) 988   (462) 1170   (463) 56   (464) 704

(465) 1650   (466) 63   (467) 76   (468) 250

(469) 779   (470) 300   (471) 966   (472) 800

(473) 252   (474) 152   (475) 510   (476) 1968

(477) 18   (478) 100   (479) 272   (480) 784

WHOLE NUMBERS

Basic Math Answer Keys

(481) 299	(482) 88	(483) 2350	(484) 1392
(485) 1536	(486) 238	(487) 135	(488) 1426
(489) 372	(490) 385	(491) 1274	(492) 0
(493) 132	(494) 1900	(495) 280	(496) 1680
(497) 288	(498) 0	(499) 100	(500) 110
(501) 0	(502) 0	(503) 490	(504) 640
(505) 1320	(506) 0	(507) 2156	(508) 660
(509) 98	(510) 3952	(511) 448	(512) 1638
(513) 0	(514) 4080	(515) 0	(516) 2142
(517) 1292	(518) 80	(519) 945	(520) 105
(521) 756	(522) 2394	(523) 630	(524) 280
(525) 544	(526) 840	(527) 1540	(528) 420

WHOLE NUMBERS

Basic Math Answer Keys

(529) 0   (530) 96   (531) 264   (532) 600

(533) 160   (534) 1512   (535) 0   (536) 56

(537) 950   (538) 5472   (539) 1960   (540) 672

(541) 6   (542) 0   (543) 0   (544) 6

(545) 13   (546) 5   (547) 15   (548) 15

(549) 9   (550) 8   (551) 3   (552) 5

(553) 13   (554) 2   (555) 12   (556) 10

(557) 11   (558) 8   (559) 15   (560) 13

(561) 14   (562) 4   (563) 2   (564) 15

(565) 10   (566) 12   (567) 3   (568) 7

(569) 2   (570) 13   (571) 10   (572) 8

(573) 6   (574) 15   (575) 4   (576) 5

WHOLE NUMBERS

Basic Math Answer Keys

(577) 8   (578) 0   (579) 6   (580) 3

(581) 2   (582) 14   (583) 11   (584) 13

(585) 0   (586) 3   (587) 3   (588) 0

(589) 4   (590) 15   (591) 3   (592) 78

(593) 34   (594) 42   (595) 35   (596) 4

(597) 49   (598) 39   (599) 5   (600) 96

(601) 12   (602) 28   (603) 49   (604) 82

(605) 10   (606) 15   (607) 20   (608) 64

(609) 84   (610) 5   (611) 48   (612) 37

(613) 37   (614) 91   (615) 47   (616) 71

(617) 27   (618) 33   (619) 58   (620) 53

(621) 91   (622) 91   (623) 95   (624) 39

WHOLE NUMBERS

Basic Math Answer Keys

(625) 65   (626) 55   (627) 46   (628) 98

(629) 59   (630) 13   (631) 92   (632) 33

(633) 28   (634) 13   (635) 62   (636) 9

(637) 56   (638) 39   (639) 11   (640) 45

(641) 44,000   (642) 142,700   (643) 370,000   (644) 92

(645) 100,000 (646) 428,760   (647) 445,090   (648) 90,000

(649) 215,580 (650) 9,000   (651) 10   (652) 18,000

(653) 32,390   (654) 5,000,000 (655) 375,871   (656) 9,500

(657) 30,000   (658) 337,000   (659) 68,400   (660) 77,689

(661) 28,400,000 (662) 3,600   (663) 86,400   (664) 61,500,000

(665) 970,000 (666) 4,350,000 (667) 2,000,000 (668) 300,000

(669) 10,000,000 (670) 900,000 (671) 350,000   (672) 90,000

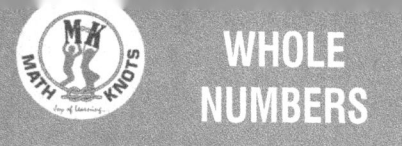

WHOLE NUMBERS

Basic Math Answer Keys

(673) 38,600,000   (674) 120,000   (675) 41,770,000   (676) 567,910,000

(677) 990,000   (678) 1,100,000   (679) 400,000   (680) 83,000,000

(681) 990,000   (682) 211,570,000   (683) 600,000   (684) 5,000,000

(685) 600,000   (686) 2,000,000   (687) 848,670,000   (688) 800,000

(689) 977,000,000   (690) 580,000,000   (691) 4,000,000,000   (692) 90,000,000

(693) 7,490,000,000   (694) 10,000,000   (695) 1,000,000,000   (696) 490,000,000

(697) 9,400,000,000   (698) 770,000,000   (699) 820,000,000   (700) 9,900,000,000

(701) 38,000,000   (702) 33,000,000   (703) 380,000,000   (704) 160,000,000

(705) 5,790,000,000   (706) 900,000,000   (707) 2,220,000,000   (708) 20,000,000

(709) 990,000,000   (710) 90,000,000   (711) 4,280,000,000   (712) 4,000,000,000

(713) 1,000,000,000   (714) 900,000,000   (715) 800,000,000   (716) 4,000,000,000

(717) 524,100,000   (718) 8,000,000   (719) 392,940,000   (720) 3,780,000

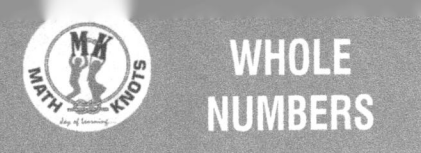

WHOLE NUMBERS

Basic Math Answer Keys

(721) 10,000,000 (722) 5,040,000 (723) 385,000,000 (724) 261,000,000

(725) 50,000 (726) 10,600,000 (727) 2,000,000,000 (728) 900,000,000

(729) 6,600,000 (730) 74,510,000 (731) 7,000,000,000 (732) 40,000,000

(733) 1,000,000,000 (734) 2,403,900 (735) 760,000 (736) 173,000,000

(737) 340,000,000 (738) 8,570,000,000 (739) 5,492,000 (740) 248,000

(741) nine hundred thirty-eight   (742) six hundred thirty-two

(743) eight hundred fifty-eight   (744) five hundred sixty-six

(745) nine hundred fifteen   (746) four hundred ninety-three

(747) one hundred fifty-four   (748) six hundred twenty-five

(750) five hundred twenty   (749) eight hundred sixty-three

(751) two hundred ninety-five   (752) nine hundred ninety-one

(753) three hundred seventy-seven   (754) one hundred twenty-six

WHOLE NUMBERS

Basic Math Answer Keys

(755) eight hundred eighty-nine

(756) seven hundred seventy-three

(757) five hundred eighty-four

(758) nine hundred forty-one

(759) one hundred fifty-three

(760) eight hundred fifty-nine

(761) five hundred seventy-one

(762) four hundred thirty-two

(763) six hundred fifty-five

(764) nine hundred seventy-one

(765) one hundred sixty-two

(766) three hundred thousand, five hundred sixty-six

(767) five hundred thousand

(768) four hundred sixty thousand, six hundred thirty-seven

(769) seven thousand, thirteen

(770) five hundred ninety thousand, seven hundred fourteen

(771) nine hundred eighty thousand, nine hundred thirty-five

(772) forty thousand, two hundred thirty-four

(773) two hundred fourteen thousand, two hundred three

WHOLE NUMBERS

Basic Math Answer Keys

(774) six hundred forty-seven thousand, two hundred sixty-nine

(775) eight hundred eighty-one thousand, eight hundred fifty-two

(776) ninety-three thousand, two hundred seventy-two

(777) twenty-five thousand, four hundred fifty

(778) nine thousand, two hundred eighty-eight

(779) one hundred eighty-one thousand, three hundred twenty-nine

(780) nine hundred sixty-four thousand, nine hundred ninety-six

(781) two thousand, nine hundred eighty

(782) three thousand, sixty-four

(783) four hundred thousand, five hundred seven

(784) seventy-four thousand, five hundred fifty-one

(785) eight thousand, four hundred eighty-two

(786) six hundred forty-one thousand, seven hundred thirty-one

(787) one hundred six thousand, three hundred ten

(788) seven thousand, eight hundred seventy-seven

(789) sixty-one thousand, six hundred fifty-five

(790) six thousand, nine hundred twenty-eight

(791) six hundred sixty-six million, seventy-one thousand, six

(792) twenty-three million, four thousand, ten

(793) four million, eight hundred seventy thousand, one

(794) five million, eight thousand, three hundred forty-five

(795) three hundred fifty-three million, four hundred three thousand, six hundred ninety

(796) eighty million, six hundred thousand, eight hundred

(797) thirty million, three hundred thousand, four hundred one

WHOLE NUMBERS

(798) three million, twenty

(799) five hundred fifty-seven million, three hundred twelve thousand

(800) three hundred thirty million

(801) one million, fifty-eight thousand, six hundred sixty

(802) thirty-six million, three hundred thousand, three hundred ninety-three

(803) forty-four million, ninety thousand, six hundred

(804) seven hundred fifteen million, twenty-five

(805) five million, eighty-nine thousand, three hundred

(806) eight hundred sixty-nine million, one hundred forty thousand, seven hundred fifty-nine

(807) eighty-eight million, twenty thousand, eighty-four

(808) eleven million, six hundred thousand, nine hundred sixty-one

(809) five million, eight hundred ninety thousand, one hundred

WHOLE NUMBERS

(810) sixty million, ninety-four thousand, nine hundred sixty

(811) sixty million, fifty thousand, ten

(812) seven million, six

(813) forty-one million, seven hundred seventy thousand, thirty

(814) seven million, four hundred nine thousand, nine hundred

(815) seven hundred three million, seven hundred fifteen

(816) seventy billion, two million, seventy thousand, eight hundred thirty

(817) four billion, five hundred sixty million, three hundred seventy-one thousand, sixty-two

(818) five hundred billion, seven hundred twenty million, three hundred thirty thousand, forty-one

(819) thirty-eight billion, one hundred million, nine hundred seventy-two thousand, one hundred forty-five

(820) seventy-three billion, five million, seven hundred thousand, five hundred

(821) six hundred seventy-six billion, five hundred ten million, six hundred eight thousand, seven hundred twenty

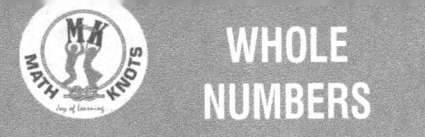

WHOLE NUMBERS

Basic Math Answer Keys

(822) five billion, one hundred sixty-five million, six hundred fifty thousand, eight hundred two

(823) two billion, six million, sixty thousand, three hundred forty

(824) six hundred billion, three million, thirty thousand, twenty

(825) forty billion, seventy-five thousand

(826) four billion, five million, six hundred thousand, one hundred six

(827) eight hundred billion, fifty-eight million, nine hundred eight thousand, four hundred

(828) two hundred fifty billion, ten million, four hundred six thousand, forty-eight

(829) nine hundred billion, five hundred ninety-five million, fifty thousand, eighty-one

(830) one billion, four hundred nine million, nine hundred thirty-six thousand, six hundred

(831) ninety billion, thirty million, ten thousand, ten

(832) sixty-four billion, one hundred million, two hundred thirty thousand

(833) ten billion, six hundred seven million, eight hundred six thousand, one hundred six

WHOLE NUMBERS

Basic Math Answer Keys

(834) fourteen billion, seven hundred fifty million, three hundred five thousand, eighty

(835) six hundred one billion, eight hundred million, five hundred eighteen thousand, five

(836) one billion, twenty million, forty thousand, nine hundred eighteen

(837) one billion, three hundred million, four hundred seven thousand, eight hundred thirty

(838) seven hundred nine billion, five hundred forty-nine million, seven hundred thirty-three

(839) fifty-three billion, one hundred one million, seven hundred five thousand, four hundred seventy

(840) nine billion, ninety-four million, eighty thousand, nine

(841) six hundred thirty trillion, sixteen billion, forty million, three hundred thousand

(842) four trillion, seventy billion, three hundred thirty million, ninety-three thousand

(843) one trillion, six hundred forty-four billion, eight million, four thousand, one hundred thirty-eight

(844) seventy-five trillion, seven hundred twenty billion, five hundred fifty million, nine hundred thousand, seven hundred fifty-four

845) three hundred fifty-four trillion, three hundred billion, one hundred two million, one thousand, five hundred fifty

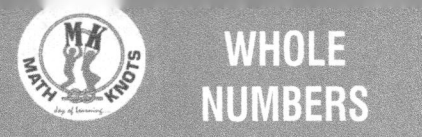

(846) one trillion, three hundred billion, one hundred two million, one thousand, five hundred fifty

(847) thirty trillion, nine billion, seventy-one million, three hundred thousand, seven hundred

(848) four hundred four trillion, two hundred billion, nine hundred sixty million, seventy-three thousand, fifty

(849) one hundred ninety trillion, seven hundred billion, eighty million, eight hundred eighty thousand, two hundred fifty-three

(850) thirteen trillion, four hundred ninety-five billion, five million, three hundred forty

(851) two trillion, four hundred billion, nine hundred seventy-seven million, one hundred thousand, thirty-six

(852) nine hundred trillion, two hundred three billion, one million, four hundred

(853) seventy trillion, two billion, nine million, five hundred thousand, two hundred two

(854) eighty trillion, three hundred ten billion, six hundred seventy-nine million, one hundred seventy thousand, five hundred five

(855) ninety-two trillion, eight hundred forty billion, nine hundred four million, one thousand, four hundred

(856) seven hundred trillion, seven billion, seven hundred thirty million, four thousand, thirty-two

(857) seven trillion, five hundred five billion, one hundred seventy-two million

WHOLE NUMBERS

Basic Math Answer Keys

(858) one trillion, three hundred eighty-two billion, three million, eighty-seven thousand, fifty

(859) four trillion, nine hundred billion, ten million, two hundred thousand, one hundred fifty-five

(860) five trillion, twenty billion, one hundred eighty million, thirty-two thousand, one hundred six

(861) eighty-one trillion, five hundred seventy-two billion, seven million, two thousand, one hundred fifty

(862) five hundred seventy-two trillion, four hundred fifty-four million, four hundred thousand, one hundred ninety

(863) six trillion, six hundred sixty billion, seven hundred two million, one hundred eighty-one thousand, seven hundred eighty-six

(864) one hundred ninety trillion, five billion, nine hundred eighty-four thousand, one hundred sixty

(865) three hundred fifty-six trillion, forty-six billion, seven hundred million, six hundred fifty thousand, thirty

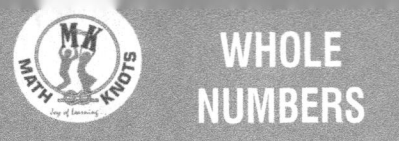

WHOLE NUMBERS

Basic Math Answer Keys

(866) 763   (867) 952   (868) 901   (869) 993

(870) 491   (871) 256   (872) 405   (873) 694

(874) 987   (875) 430   (876) 817   (877) 526

(878) 608   (879) 641   (880) 244   (881) 303

(882) 989   (883) 357   (884) 848   (885) 560

(886) 806   (887) 297   (888) 258   (889) 172

(890) 463   (891) 909,090   (892) 5,207   (893) 9,506

(894) 70,006   (895) 947,701   (896) 800,030   (897) 21,090

(898) 600,700   (899) 206,340   (900) 7,600   (901) 80,600

(902) 500,790   (903) 343,854   (904) 9,600   (905) 6,120

(906) 8,700   (907) 20,119   (908) 79,030   (909) 17,800

(910) 650,165   (911) 3,400   (912) 90,764   (913) 53,500

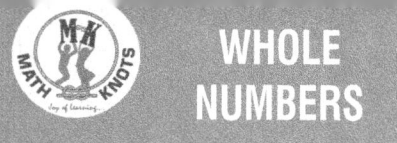

WHOLE NUMBERS

Basic Math Answer Keys

(914) 10,060   (915) 92,200   (916) 48,187,002   (917) 303,800,000

(918) 560,030,009   (919) 7,035,900   (920) 6,900,709   (921) 1,045,998

(922) 6,001,090   (923) 2,206,005   (924) 3,096,006   (925) 1,000,400

(926) 6,828,006   (927) 10,515,300   (928) 69,531,000   (929) 74,900,090

(930) 260,702,026   (931) 44,462,048   (932) 2,862,006   (933) 1,090,004

(934) 9,107,600   (935) 97,572,606   (936) 131,709,603   (937) 4,090,002

(938) 309,120,090   (939) 5,308,109   (940) 85,400,040   (941) 70,000,090,194

(942) 491,862,000,000   (943) 28,410,589,680   (944) 3,600,066,510   (945) 81,000,002,134

(946) 963,850,403,001   (947) 36,008,601,056   (948) 20,006,600,012   (949) 70,870,200,881

(950) 2,134,009,869   (951) 580,795,523,019   (952) 106,123,306,805   (953) 78,000,900,070

(954) 700,300,705,033   (955) 7,200,863,063   (956) 43,060,398,890   (957) 2,141,200,509

(958) 69,098,707,403   (959) 40,580,906,000   (960) 9,000,508,391   (961) 41,161,880,768

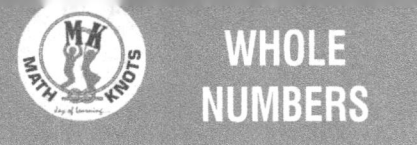

WHOLE NUMBERS

Basic Math Answer Keys

(962) 3,004,123,346   (963) 80,103,500,795   (964) 713,230,700,008   (965) 8,098,000,700

(966) 530,090,919,005,000   (967) 88,659,602,540,020   (968) 4,090,062,666,000

(969) 802,770,208,775,000   (970) 6,060,744,580,432   (971) 70,008,180,870,001

(972) 11,716,105,002,008   (973) 20,194,008,530,060   (974) 16,034,005,706,370

(975) 25,200,001,440,400   (976) 869,021,020,932,005   (977) 6,301,076,107,200

(978) 10,002,360,710,502   (979) 83,042,400,600,360   (980) 40,350,064,000,712

(981) 190,002,780,091,790   (982) 9,320,002,780,089   (983) 10,000,080,004,030

(984) 6,010,070,933,505   (985) 500,590,375,065,032   (986) 20,037,083,601,200

(987) 1,035,300,430,000   (988) 97,640,512,000,800   (989) 40,607,095,040,000

(990) 3,004,000,007,701   (991) thousands   (992) ten thousands

(993) hundreds   (994) ten thousands   (995) thousands   (996) hundreds

(997) ten thousands   (998) ten thousands   (999) thousands   (1000) ten thousands

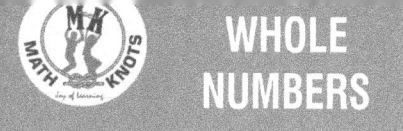

WHOLE NUMBERS

Basic Math Answer Keys

(1001) ten thousands   (1002) thousands   (1003) hundreds   (1004) thousands

(1005) thousands   (1006) thousands   (1007) ten thousands   (1008) hundreds

(1009) ten thousands   (1010) ten thousands   (1011) hundreds   (1012) thousands

(1013) hundreds   (1014) ten thousands   (1015) ten thousands   (1016) hundreds

(1017) thousands   (1018) thousands   (1019) thousands   (1020) ten thousands

(1021) ten thousands   (1022) thousands   (1023) hundreds   (1024) hundreds

(1025) thousands   (1026) hundreds   (1027) hundreds   (1028) ten thousands

(1029) ten thousands   (1030) hundreds   (1031) hundreds   (1032) hundreds

(1033) thousands   (1034) thousands   (1035) ten thousands   (1036) hundreds

(1037) thousands   (1038) thousands   (1039) thousands   (1040) ten thousands

(1041) ten millions   (1042) ten millions   (1043) ten millions   (1044) millions

(1045) millions   (1046) millions   (1047) ten millions   (1048) ten millions

WHOLE NUMBERS

Basic Math Answer Keys

(1049) hundred thousands   (1050) ten millions   (1051) ten millions

(1052) hundred thousands   (1053) millions   (1054) millions

(1055) ten millions   (1056) hundred thousands   (1057) hundred thousands

(1058) millions   (1059) ten millions   (1060) hundred thousands

(1061) hundred thousands   (1062) ten millions   (1063) millions

(1064) hundred thousands   (1065) millions   (1066) millions

(1067) ten millions   (1068) millions   (1069) hundred thousands

(1070) millions   (1071) ten millions   (1072) ten millions

(1073) hundred thousands   (1074) ten millions   (1075) ten millions

(1076) millions   (1077) hundred thousands   (1078) ten millions

(1079) hundred thousands   (1080) hundred thousands   (1081) millions

(1082) millions   (1083) ten millions   (1084) ten millions   (1085) ten millions

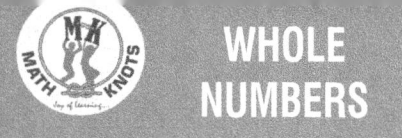

WHOLE NUMBERS

Basic Math Answer Keys

(1086) ten millions (1087) hundred thousands (1088) ten millions

(1089) hundred thousands (1090) ten millions (1091) hundred millions

(1092) hundred millions (1093) hundred millions (1094) billions

(1095) ten millions (1096) ten millions (1097) billions (1098) ten millions

(1099) ten millions (1100) ten millions (1101) billions (1102) ten millions

(1103) hundred millions (1104) billions (1105) ten millions

(1106) ten millions (1107) ten millions (1108) billions (1109) ten millions

(1110) billions (1111) billions (1112) billions (1113) ten millions

(1114) billions (1115) ten millions (1116) hundred millions

(1117) billions (1118) billions (1119) billions (1120) ten millions

(1121) hundred millions (1122) hundred millions (1123) hundred millions

(1124) ten millions (1125) billions (1126) ten millions (1127) ten millions

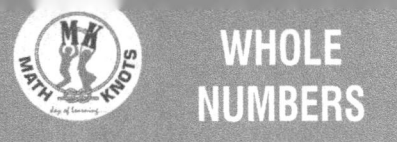

WHOLE NUMBERS

Basic Math Answer Keys

(1128) billions (1129) billions (1130) ten millions (1131) ten millions

(1132) hundred millions (1133) hundred millions (1134) ten millions

(1135) hundred millions (1136) hundred millions (1137) ten millions

(1138) hundred millions (1139) ten millions (1140) hundred millions

(1141) 402,000 (1142) 12,548,000 (1143) 50,000 (1144) 321,000

(1145) 25,437,000 (1146) 39,321,000 (1147) 366,000 (1148) 990,000

(1149) 783,000 (1150) 190,000 (1151) 7,200 (1152) 92,490,000

(1153) 837,300 (1154) 273,000 (1155) 81,660,000 (1156) 705,800

(1157) 597,000 (1158) 7,380,000 (1159) 35,000 (1160) 878,010,000

(1161) 52,831,700 (1162) 19,811,000 (1163) 6,290,000 (1164) 2,500

(1165) 8,499,000 (1166) 250,000 (1167) 589,700 (1168) 369,300

(1169) 57,000 (1170) 620,900 (1171) 900 (1172) 6,000

WHOLE NUMBERS

Basic Math Answer Keys

(1173) 5,000　　(1174) 8,320,000　　(1175) 3,000　　(1176) 9,000

(1177) 25,853,000　　(1178) 67,690,000　　(1179) 197,100　　(1180) 4,750,000

(1181) 725,900　　(1182) 370,000　　(1183) 7,740,000　　(1184) 85,282,800

(1185) 4,900　　(1186) 9,000　　(1187) 30,000　　(1188) 67,229,700

(1189) 95,340,000　　(1190) 405,270,000　　(1191) 72.616780　　(1192) 32.786

(1193) 30.032883　　(1194) 5,000,000,000　　(1195) 28.198770　　(1196) 0.9686

(1197) 97.7307　　(1198) 1,000,000,000　　(1199) 6,500,000　　(1200) 0.9498

(1201) 5.633557　　(1202) 1,000,000,000　　(1203) 900,000,000　　(1204) 549,000,000

(1205) 400,000,000　　(1206) 31,000,000　　(1207) 1,000,000,000　　(1208) 2,000,000,000

(1209) 2,000,000,000　　(1210) 700,000　　(1211) 3,000,000,000　　(1212) 88.102

(1213) 6.973　　(1214) 6.5500　　(1215) 900,000　　(1216) 9.658626

(1217) 210,000,000　　(1218) 17.263558　　(1219) 0.926　　(1220) 9,000,000

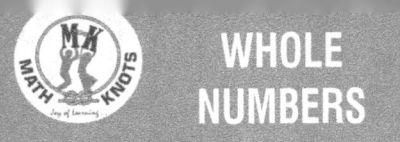

WHOLE NUMBERS

Basic Math Answer Keys

(1221) 6.7477   (1222) 65,000,000   (1223) 2,000,000   (1224) 9,000,000,000

(1225) 5,000,000   (1226) 100,000,000   (1227) 690,000,000   (1228) 500,000,000

(1229) 3.2899   (1230) 100,000,000   (1231) 200,000,000   (1232) 100,000

(1233) 1,300,000,000 (1234) 36.3173   (1235) 9.9110   (1236) 30,000,000

(1237) 200,000,000   (1238) 5.9863   (1239) 2.279   (1240) 80,000,000

(1241) 2.699   (1242) 9.949310   (1243) 8.2295   (1244) 9,000,000

(1245) 1.9473   (1246) 400,000   (1247) 52.97669   (1248) 21.398

(1249) 7.524505   (1250) 4.558072   (1251) 67,500,000   (1252) 19.221

(1253) 800,000   (1254) 2,000,000,000   (1255) 2,000,000,000   (1256) 7.362521

(1257) 7,000,000,000 (1258) 10,000,000,000   (1259) 9,000,000,000   (1260) 6.480

(1261) 9,000,000,000 (1262) 3.508318   (1263) 6.44250   (1264) 629,000,000

(1265) 10,000,000,000

www.ingramcontent.com/pod-product-compliance
Lightning Source LLC
Chambersburg PA
CBHW081744100526
44592CB00015B/2289